FIELD GUIDE TO THE

NEIGHBORHOOD BIRDS

OF NEW YORK CITY

ALSO BY LESLIE DAY

FIELD GUIDE TO THE NATURAL WORLD OF NEW YORK CITY

FIELD GUIDE TO THE STREET TREES OF NEW YORK CITY

FIELD GUIDE TO THE
NEIGHBORHOOD BIRDS
OF NEW YORK CITY

LESLIE DAY

ILLUSTRATED BY **TRUDY SMOKE**
PHOTOGRAPHS BY **BETH BERGMAN**
FOREWORD BY **DON RIEPE**

JOHNS HOPKINS UNIVERSITY PRESS
BALTIMORE

© 2015 Johns Hopkins University Press
All rights reserved. Published 2015
Printed in China on acid-free paper
9 8 7 6 5 4 3 2

Johns Hopkins University Press
2715 North Charles Street
Baltimore, Maryland 21218-4363
www.press.jhu.edu

Library of Congress Cataloging-in-Publication Data

Day, Leslie, 1945– author.
  Field guide to the neighborhood birds of New York City / Leslie Day ; illustrated by Trudy Smoke;
  photographs by Beth Bergman; foreword by Don Riepe
    pages cm
  Includes index.
  ISBN 978-1-4214-1617-5 (hardcover : alk. paper) — ISBN 1-4214-1617-4 (hardcover : alk.
  paper) — ISBN 978-1-4214-1618-2 (pbk. : alk. paper) — ISBN 1-4214-1618-2 (pbk. : alk. paper)
  — ISBN 978-1-4214-1619-9 (electronic) — ISBN 1-4214-1619-0 (electronic) 1. Birds—New
  York (State)—New York—Guidebooks. 2. New York (N.Y.)—Guidebooks. I. Smoke, Trudy. II.
  Bergman, Beth. III. Title.
  QL684.N7D39 2015
  598.09747′1—dc23   2014024716

A catalog record for this book is available from the British Library.

*Special discounts are available for bulk purchases of this book. For more information, please
contact Special Sales at 410-516-6936 or specialsales@press.jhu.edu.*

Johns Hopkins University Press uses environmentally friendly book materials, including recycled
text paper that is composed of at least 30 percent post-consumer waste, whenever possible.

Book design by Kimberly Glyder

CONTENTS

FOREWORD

M ost people think of New York City as a concrete jungle devoid of any bird life save pigeons, starlings, and house sparrows. Actually, quite the opposite is true as Leslie Day's *Field Guide to the Neighborhood Birds of New York City* reveals. A quick thumb-through of her book brings to light the range and diversity of bird species found within the city's five boroughs. Nestled between the Atlantic Ocean, Hudson River, and Long Island Sound, the city supports a great wealth of wildlife, especially birds. The many miles of shoreline and variety of upland parks provide ample nesting, wintering, and migratory stopover habitat for many species of shorebirds, waterfowl, and songbirds. Three hundred forty species of birds have been recorded at the Jamaica Bay Wildlife Refuge alone. Located along the Atlantic migratory flyway in the boroughs of Brooklyn and Queens, this 9,155 acre preserve is internationally renowned as a significant urban hotspot for birds and deemed a Critical Fish and Wildlife Habitat by New York State and an Important Bird Area by the National Audubon Society. The refuge harbors many unusual and colorful bird species, such as the glossy ibis, American oystercatcher, eastern towhee, northern shoveler, and brant goose. Elsewhere, the city's parks, bridges, and buildings are home to nesting red-tailed hawks, great horned owls, American kestrels, and peregrine falcons.

The five boroughs of the city provide habitats that range from mature upland woods, shrub thickets, open grasslands, and marshes, to miles of shoreline, sandy beaches, and open water. In spring and fall, literally millions of birds fly through and above the city during their migrations from wintering to breeding grounds and back. Many of the area's premier parks, including Central Park (Manhattan), Forest Park (Queens), Prospect Park (Brooklyn), Mount Loretto Nature Preserve (Staten Island), and Pelham Bay Park (Bronx), are considered some of the best birding sites in the northeastern United States. These parks are oases of green habitat for large concentrations of warblers, tanagers, grosbeaks, and many other colorful birds as they wing their way north and south each year.

This informative guide is made beautiful by the illustrations of Trudy Smoke and the amazing photographs of Beth Bergman and other local photographers. The writing and imagery capture the essence of the avian life that surrounds us. We are shown how to identify the birds, how they behave, and where and when we can see them. Leslie Day includes information about the many groups that work tirelessly to protect New York City's birds. Everything you need to know about our birds and birding in New York City can be found in this portable book. This field guide will help beginning bird-watchers find their way, but it also belongs beside the chair near the feeder and in the library of every serious birder between the Rockaways and Riverdale.

Don Riepe
Jamaica Bay Guardian
American Littoral Society

ACKNOWLEDGMENTS

The beauty of birds as they fly, feed, bathe, care for one another, and raise their young has been captured by the photographers and artists who shared their work with me. Two extraordinary women are foremost among them: my lifelong friend, the talented illustrator Trudy Smoke, whose day job is linguistics professor at Hunter College CUNY, where she has taught for more than forty years, and Beth Bergman, who, when she is not photographing the Metropolitan Opera Company, which she has done for more than forty years, is a passionate avian photographer. I am deeply indebted to the following for their spectacular photographs: Don Riepe, Guardian of the Jamaica Bay Wildlife Refuge; Jenny Mastantuono, wildlife biologist, USDA Wildlife Services; Laura Meyers, New York City bird photographer; David Goldemberg; Marge Pangione, Sammie Smith, and bird bloggers James O'Brien and D. Bruce Yolton.

I could not have done without the help of Glenn Phillips and his encyclopedic knowledge of birds. As former Executive Director of New York City Audubon, Glenn supported me in countless ways, among them, giving me access to his hard-working and talented staff: Susan Elbin, ornithologist and Director of Conservation and Science; Tod Winston, communications manager, research associate, and one of the most accurate bird identification people I know; and Kaitlyn Parkins, a researcher on New York City conservation issues.

My everlasting gratitude goes to Rob Jett, the City Birder, for fact checking all my species' accounts. He is the fastest, most accurate fact-checker around! Many, many thanks to the following for checking the neighborhoods that surround the birding hotspots of the five boroughs: Linda Dallam for reviewing the Bronx neighborhoods; Andrew Baksh and Don Riepe for checking the Queens neighborhoods; John Kilcullen, Conference House Park Director, for checking the Staten Island neighborhoods; and Rob Jett for checking the Brooklyn neighborhoods.

My eternal gratitude to New York City wildlife rehabilitators Ritamary McMahon, founder of the Wild Bird Fund, and Cathy and Bobby Horvath, founders of Wildlife in Need of Rescue and Rehabilitation, for their lifetime devotion to hurt and orphaned animals and for their patience in answering my many questions. I am grateful to New York City Urban Park Ranger Rob Mastrianni for the work he and other rangers are doing to bring injured wildlife to those who can help them survive.

A huge thank-you to my loving husband, Jim Nishiura, for his continued support through each of my books and to my darling son, Jonah, and his precious wife, Gina, for always being there for me.

To my colleagues at The Elisabeth Morrow School, particularly Nancy Dorrien, Gil Marino, Gail Weeks, and Sammie Smith, a huge thank-you for your love and support.

Heartfelt gratitude to my friends at Johns Hopkins University Press, particularly my talented publicist, Kathy Alexander, and highly skilled editor, Andre Barnett. Finally, my love and gratitude to Vincent J. Burke, my brilliant, patient, funny, and ingenious editor at Johns Hopkins, for giving me the opportunities to write these books on the natural world of New York City, my home and the place where my heart and soul reside.

FOR MY NEST-MATE, JIM

INTRODUCTION

I n New York City, birds are everywhere, but for the first 38 years of my life, I could only name a few of them: pigeons, starlings, robins, and the little brown birds I collectively called sparrows. Even though I was drawn to the natural world, and in 1975 moved into a houseboat on the Hudson River at the West 79th Street Boat Basin, I was uneducated about the nature all around me. Until 1983.

One morning I walked our two dogs around the ball field next to the Boat Basin, scattering leftover seed from my neighbors' parrot, Bobo. Suddenly a gorgeous, unfamiliar little bird flew up to me. She was about 8 inches long, tawny brown, with a red crest and a red beak. She followed me around the ball field and ate the seed I scattered. She awoke something inside me, and I had to know what kind of bird she was. I borrowed a field guide to birds from a friend and found her: a female northern cardinal.

Woman feeding pigeons on park bench.

For the next three years, she accompanied me on my dog walks every day and in every season. She would sit on the railing along the river and start calling to me before sun-up. Tsip! Tsip! During a blizzard, she flew to me as I rounded the traffic circle on my walk home from the 92nd Street Y Nursery School, where I taught at the time. I ran to the boat, got sunflower seeds, and ran back to the park where she, along with blue jays, house sparrows, starlings, and pigeons

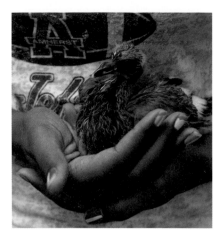

Young volunteer holding baby pigeon at Wildbird Fund.

waited for food. I cleared the snow away and scattered the seed and the birds flew to the ground, led by my amazing cardinal. After three years, a male cardinal showed up, and she flew away with him. I never saw her again, but she changed my life. I wanted to know the name of every bird that lived in or flew through my city.

I now know what every birder knows: Once you connect with birds, you will see them everywhere. They share the sidewalks with us. They build their nests on, above, and below the ledges of our apartments, brownstones, office towers, and bridge spans. They sometimes devour the pizza slice on the ground or the seed scattered by a neighbor, but they also consume the cornucopia of berries, flowers, and seeds of our city streets, backyards, and parks. Along our coastline they nest, raise their young, and feed on the bounty produced by the sea and streams that surround and flow through our city, consuming fish, clams, oysters and mussels, and the nutritious sea grass that struggles to survive along our beaches.

New York City's birds receive a lot of help. Every day, birds are injured by flying into our skyscrapers or are sickened when they feed on toxins we introduce into their environment. Luckily, there are organizations that care for these injured animals. These groups take in hurt and sick birds and rehabilitate and release them back into the wild. Rita McMahon and the Wild Bird Fund she created is one example. When I visited the Wild Bird

Fund recently, a Canada goose swam in their tank. He had wandered into the Brooklyn Battery Tunnel and was hit by two cars. The tunnel police scooped him up and drove him to the Wild Bird Fund, where they let him heal from bad bruises, eventually releasing him in the North Cove in the Harlem River.

Bobby and Cathy Horvath and their organization Wildlife in Need of Rescue and Rehabilitation are the go-to rescuers for birds and mammals, particularly birds of prey. Bobby, a New York City firefighter, and Cathy, a licensed vet tech, care for more than 700 injured and sick animals a year and bring many of them to educational events throughout the city, teaching generations of New Yorkers about the needs of our raptors: eagles, hawks, falcons, and owls.

New York City Audubon has many beneficial programs for birds. Former Executive Director Glenn Phillips talked with me about the importance of protecting their habitat by planting native shrubs and trees that produce berries and seeds they feed on and that attract the insects they prey on. A major effort is working to make skyscrapers less dangerous when birds migrate at night. Confused by the lights, nearly 90,000 birds

Bobby and Cathy Horvath holding a Eurasian eagle-owl and a rescued northern saw-whet owl at an educational event for the Fort Tryon Park Trust in northern Manhattan.

Colonial nesting great egrets, herring gulls, and double-crested cormorants on an island in Jamaica Bay. *JM*

die annually by colliding with city buildings at night. In 2005, New York City Audubon Society launched the Lights Out New York Program, and since then, the Chrysler Building, the Empire State Building, Rockefeller Center, the Time Warner Center, and other buildings dim their lights from midnight until dawn during the height of fall migration, September 1 through November 1. This effort keeps thousands of birds safe as they migrate through Manhattan. New York City Audubon also works toward educating New Yorkers about the terrible toll outdoor cats take on wild birds, encouraging residents to keep their cats indoors to protect our bird life. Susan Elbin, the Society's Director of Conservation and Science manages the organization's Harbor Herons Project, which monitors and protects the eggs, nests, and habitat of gulls, wading birds, and shorebirds that nest on the city's small islands.

There is a great history of interest in and caring for birds in New York City. John James Audubon, who devoted his life to creating life-sized paintings of birds, traveled North America in the early nineteenth century to paint every bird species. Audubon spent the last years of his life in a house on West 155th Street along the banks of the Hudson River. He is buried in the Trinity Church cemetery where his tomb, the Audubon monument, stands on West 155th Street. The area, in what is now Washington Heights, is known as Audubon Park. Many of Audubon's original paintings hang in the New York Historical Society, which owns more of his work than any other institution in the world.

Other New York bird connections include Frank Chapman, chief ornithologist and curator of birds at the American Museum of Natural History on West 81st Street and Central Park West, at the turn of the nineteenth century. He designed the museum's famous dioramas that showed birds and their habitats, including species on Florida's Pelican Island that were on the verge of extinction. In 1903, Chapman's diorama inspired President Theodore Roosevelt, whose father was a trustee of the museum, to make Pelican Island the first National Wildlife Preserve.

Roger Tory Peterson, the noted bird artist and creator of the Peterson Field Guide to Birds, started his career as a founding member of the Bronx County Bird Club. He explored the rich bird life of the city in the 1920s, combing the five boroughs' parks, sewer outfalls, and garbage dumps for birds. He once marveled about his sighting of four snowy owls, which were feeding on rats at the Hunts Point dump.

Thousands of New Yorkers are passionate about watching birds, and they are always ready to teach someone new to the adventure. In Central Park, during the height of spring migration, more than 500 birders may gather on a single day. Birders, typically kind and helpful, will let you or your children look through their scopes at red-tailed hawks in their nest, tenderly feeding their young.

Once you start to notice birds, you will see them everywhere: outside your classroom, your office, the stores you shop in, your hospital window. This guide will help you identify them, learn about them, and learn from them what it is we can do to help them survive. Take this guide wherever you go and enter the enchanting and interesting world of birds. New York City will never look the same.

FIELD GUIDE TO THE

NEIGHBORHOOD BIRDS

OF NEW YORK CITY

ILLUSTRATED

BIRD, WING,

---

and FEATHER

ANATOMY

ANATOMY OF A BIRD

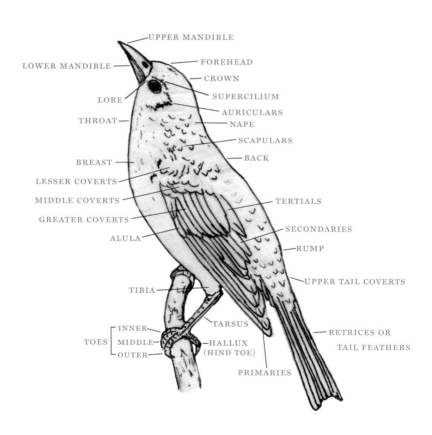

UPPER MANDIBLE

LOWER MANDIBLE

FOREHEAD

CROWN

LORE

SUPERCILIUM

THROAT

AURICULARS

NAPE

SCAPULARS

BREAST

BACK

LESSER COVERTS

MIDDLE COVERTS

TERTIALS

GREATER COVERTS

SECONDARIES

ALULA

RUMP

UPPER TAIL COVERTS

TIBIA

INNER

TARSUS

TOES | MIDDLE

RETRICES OR

OUTER

HALLUX
(HIND TOE)

TAIL FEATHERS

PRIMARIES

ANATOMY OF A BIRD WING

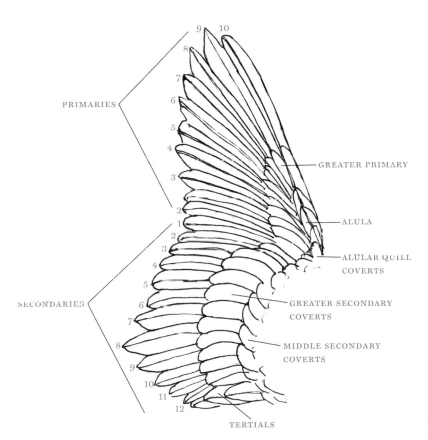

PRIMARIES

9 10
8
7
6
5
4
3
2
1

GREATER PRIMARY

ALULA

ALULAR QUILL
COVERTS

SECONDARIES

2
3
4
5
6
7
8
9
10
11
12

GREATER SECONDARY
COVERTS

MIDDLE SECONDARY
COVERTS

TERTIALS

ANATOMY OF A FEATHER

FLIGHT FEATHER (REMEX)

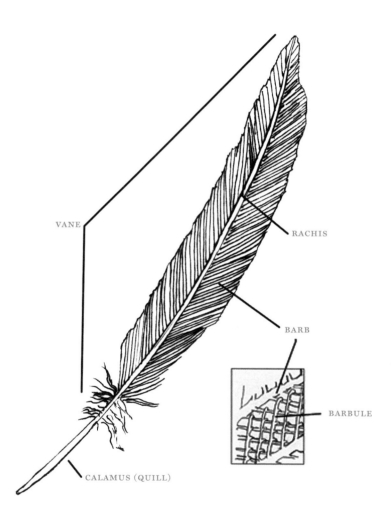

VANE

RACHIS

BARB

BARBULE

CALAMUS (QUILL)

BIRD

TERMINOLOGY

Allopreening: Mutual preening, when birds preen the feathers of another.

Altricial: Young birds when hatched, that have no feathers, closed eyes, and are totally dependent on their parents.

Alula: A small group of stiff feathers on the "thumb bone" of a bird's wing.

Anting: Birds spread out their wings on active ants and as the ants climb all over them, they release formic acid on the birds' feathers, which most likely kills parasites and repels insects.

Auriculars: Ear coverts or ear patches.

Beard: Long feathers hanging from a male turkey's breast. Ten percent to 20 percent of female turkeys have a beard.

Breast: Area between the throat and the belly.

Breeding plumage: Colorful feathers on adult birds during courtship.

Brood: Baby birds.

Brood parasite: Birds that lay their eggs in the nests of other birds to be raised by the other birds.

Carpal bar: Dark shoulder bar.

Cere: Waxy, fleshy area at the base of the upper mandible in some species that houses the nostrils.

Clutch: All the eggs in the nest incubated by the female.

Colonial: Birds that build nests next to one another in colonies.

Contour feathers: Feathers, controlled by individual muscles, that cover most of the bird, keeping it dry and warm.

Countersinging: One bird sings in response to another bird's song.

Coverts: Smaller feathers that cover the wings.

Crest: Feathers that stand up on the crown of a bird's head. Many crests can be raised and lowered.

Crop: The area where food is stored to be digested later.

Crown: Top of a bird's head.

Culmen: Top ridge line of a bird's upper bill.

Dabbling: A method of feeding by some ducks and all geese and swans on the surface of the water or by tipping over with their legs in the air and their bills underwater to reach food below the surface.

Decurved: Bills that curve downward.

Diurnal: Birds that are active during daylight hours.

Ear coverts: Feathers behind the eyes that cover the ears.

Eclipse plumage: Dull colors of feathers during nonbreeding seasons.

Eye crescents: White crescents above and below the eye.

Eye ring: A line circling the bird's eye.

Fledge: When a young bird leaves its nest.

Fledgling: Young bird that has left the nest but is still dependent on its parents for food and protection.

Flight feathers: Large wing and tail feathers.

Gape: The inside of a bird's open mouth. On young birds, the gape is the brightly colored flange where the upper and lower mandible meet.

Gonydeal spot: Red dot on the lower mandible of an adult gull that, when pecked by its chick, stimulates it to feed its young.

Gonys: Bottom edge of the lower mandible of a gull.

Gorget: Iridescent throat feathers on male, and some female, hummingbirds.

Gosling: Young goose.

Hallux: Hind toe.

Hawking: Catching insects on the wing.

Juvenile: Immature bird.

Keratin: Protein that makes up beaks, feathers, and talons.

Lamellae: Tiny, toothlike ridges along the cutting edges of a waterfowl's bill that filters food from the water.

Lore: Area between the bill and the eye.

Malar: Cheek.

Mandibles: Lower and upper bills.

Mantle: Bird's back, shoulders, upperwing-coverts, and secondary feathers.

Maxilla: Upper bill.

Molt: Growing new feathers after the old or damaged feathers have been shed.

Morph: Color variation in the same species of a bird. The screech owl has gray and red color morphs.

Nape: Back of the head.

Niche: Carnivore, omnivore, herbivore, insectivore, or pollinator role played by bird species within its ecological community.

Nocturnal: Birds that are active at night.

Nonpasserines: Nonperching birds, such as shorebirds, waterfowl, birds of prey, pigeons, and doves.

Orbital ring: Unfeathered, naked skin around the eye.

Passerines: More than half of all bird species that have three toes forward and one back, which allows them to perch; often called songbirds.

Pip: Hatchling breaks through its shell.

Precocial: Birds that hatch with their eyes open, down-covered feathers, and the ability to leave the nest within a few days.

Preening: The act of feather maintenance, including waterproofing by spreading oil over the feathers from the preen gland, located at the base of the bird's lower back near its tail.

Primaries: Longest wing feathers of the outer wing.

Raptor: Bird of prey.

Rectrices: Tail feathers.

Remiges: Flight feathers of the wing.

Rictal bristles: Surround the bills of many insect-eating birds and are thought to protect their eyes as they feed, to hold their prey in place, and to provide sensory feedback.

Roost: Resting site.

Scapulars: Feathers at the top of a bird's wing.

Secondaries: Flight feathers of the wing closer to the bird's body than the primaries.

Semialtricial: Young birds hatched with eyes open, down-covered feathers, but without the ability to leave their nest.

Semiprecocial: Young birds hatched with eyes open, down-covered feathers, but continue to stay in the nest.

Sexual dimorphism: Male and the female of the same species look different.

Speculum: Colorful feather patches on the wings of ducks.

Stoop: Fast downward flight of a bird after prey.

Subterminal band: Located at the end of the tail feathers.

Subterminal spots: Located on the tip of the primary feathers.

Supercilium: Eyebrow-like line over bird's eye.

Talons: Claws of a bird of prey.

Tarsi: Long foot bones that lead into the bird's toes.

Tertials: Flight feathers at the base of the wing, closest to the body.

Tibia: Upper leg bone.

Undertail coverts: Small feathers covering the undertail.

Underwing coverts: Small feathers covering the base of the underwing of the bird.

Uppertail coverts: Small feather covering the base of the upper side of the tail feathers and rump.

Wattle: Fleshy skin that hangs from the lower bill of some species, like the wild turkey.

Wing bar: Rows of color patterns on the flight feathers of a bird's wing, which aids in species identification.

Wing coverts: Feathers that cover the base of the flight feathers on a bird's wing.

Zygodactyl: Two toes that face forward and two toes that face backward for clinging to tree trunks. Typically found on woodpeckers.

BIRDS

CORMORANTS belong to the family Phalacrocoracidae of which there are 40 species of cormorants worldwide. Common characteristics are brightly colored featherless skin on the lores and gular pouch, which can be blue, orange, red, or yellow, turning more brightly colored in the breeding season. These seabirds are typically black. The word *cormorant* means sea raven in Latin. Their feet are four-toed and webbed. Cormorant species live throughout the world, nesting along coastlines and freshwater habitats. They are all diving birds, consuming benthic fish, most often in shallow water. Their hooked bills help them attack their prey, and their expandable gular pouches allow them to hold their food *inside* their mouths when the fish struggle.

Using cormorants to fish is an ancient practice, which still occurs in parts of China. In Japan, it is called *Ukai* and has been used for 1,300 years by Japanese fishermen. It still goes on as a tourist attraction on several rivers in Japan during the summer. Fishermen have six or more sea cormorants on long tethers that dive under the water to catch fish. A rope tied around their necks prevent them from swallowing the fish.

The double-crested cormorant is native to North America and commonly found in New York City. The double crests are visible during breeding season.

Cormorants are an ancient and interesting bird whose ancestors go back to the time of the dinosaurs.

DOUBLE-CRESTED CORMORANT

Egg shown at life size.

Double-crested Cormorant: *Phalacrocorax auritus*

Where and when to find During spring, summer, and early fall throughout the city's rivers, lakes, ponds, sound, bays, and ocean, typically seen standing atop pilings holding their wings out to dry or swimming low in the water before they dive for fish. Their numbers increased greatly during the 1980s and '90s, but have now leveled off. Most cormorants migrate south in the fall so they can continue to fish.

What's in a name? *Phalacrocorax*: bald raven; *auritus*: crested, referring to the rarely seen double-crested ear tufts. *Cormorant*: sea raven.

Description Large black body with a long black tail, black legs and feet; bare orange throat patch and lores and turquoise-green eyes. Double-crest feathers appear during breeding season. Immatures have a light-brown throat and chest.

Size 29–36 inches long; wingspan: 54 inches.

Behavior One of the few birds that do not have oil in their feathers. Wet feathers are an advantage because cormorants dive and stay submerged to find fish. Sealed nostrils, a long rudder-like tail, and superb underwater vision assist their fishing. As they perch on pilings, cormorants hold their wings out to dry. It is easy to spot cormorants as they fly in groups very low over the water or swim low in the water, dive, and then resurface.

Nest and eggs Commonly nest in the city. Parents use sticks and material like rope, old balloons, fishing net, and plastic, which they weave into the flat nest. Eggs are almost white. During the hot summer, the parents shade the chicks from the sun, bringing them both food and water. U Thant Island, which lies in the East River across from the United Nations and was named for the former Secretary General, has many cormorant nests.

Voice Usually silent but can make deep grunts.

Ecological role Cormorants are carnivores and can dive to 25 feet and stay submerged for several minutes as they hunt for fish. In freshwater, they also consume amphibians, crayfish, and aquatic invertebrates.

Breeding double-crested cormorant showing crests, Green-Wood Cemetery, Sylvan Water pond, April. *LM*

Cormorants and their stick nests cover a tree on an island in Jamaica Bay. *JM*

Cormorant hatchlings in nest high in tree showing their gular pouches. *JM*

Two double-crested cormorants on rock at the Lake, Central Park.

Adult double-crested cormorant holding its wings out
to dry on perch over the Lake, Central Park.

Double-crested cormorant with bluegill sunfish in the Lake, Central Park.

DUCKS, GEESE, AND SWANS make up the Anatidae family of birds, with almost 150 species worldwide and 64 species in North America. The evolutionary characteristics that they share include highly water-proofed feathers; a spatulate-shaped bill with a horny tip called a nail for rooting for food in the muck; a large, round body with webbed feet toward the back. The first three toes are webbed, making it easy for the birds to move through water. They have lamellae along the sides of their bills, which help them filter the water for food: mostly vegetation and sometimes snails, tadpoles, and aquatic insects. Most species within this bird family are monogamous, either for the breeding season as with most ducks or for life as with the Canada goose and the mute swan.

New York City is awash in ducks, geese, and swans. Some species live here year-round, nesting along our lakes, ponds, and rivers, but many species, such as the brant goose, wood duck, buffleheads, and mergansers, spend winters on our wetlands when the waters in their northern breed-ing territories are frozen over.

During the summer, it is easy to spot mallards, Canada geese, and mute swans on their nests and leading their ducklings, goslings, and cygnets to water. During a winter visit to Jamaica Bay Wildlife Refuge you will see the great spectacle of thousands of brant geese and snow geese wheeling over the bay and hundreds of wintering ruddies, shovelers, and buffleheads on the ponds.

BRANT
GOOSE

CANADA
GOOSE

MUTE
SWAN

Eggs shown at ⅓ life size.

WOOD DUCK

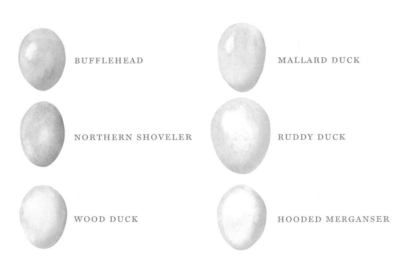

BUFFLEHEAD

MALLARD DUCK

NORTHERN SHOVELER

RUDDY DUCK

WOOD DUCK

HOODED MERGANSER

Eggs shown at ⅓ life size.

Brant Goose: *Branta bernicla*

Where and when to find Winter along the city's shores. They can be seen flying in huge flocks over the salt marshes of Jamaica Bay in Brooklyn and Queens and in Wolfe's Pond Park on Staten Island. New York City environmentalist Glenn Phillips, calls this bird a "responsibility species." Eighty thousand birds, the entire Atlantic population, use Jamaica Bay Wildlife Refuge as a staging area in spring before their migration to their Arctic breeding grounds. They depend on the refuge for their survival.

What's in a name? *Branta:* charcoal coloration; *bernicula:* from the Irish legend that these geese hatched from barnacles.

Description Adults have black heads, breasts, and necks with a white necklace; brown-barred white flank and belly feathers. Their small bills and webbed feet are black. Primary feathers, flight feathers, and tail are also black, but upper tail coverts are bright white, giving them a white-rumped appearance in flight.

Size 22–26 inches long; wingspan: 42 inches.

Behavior: Their migration is one of the longest of any waterfowl. From their nests along the Arctic coast, they can travel 3,000 miles to their wintering feeding grounds. Some fly 7,000 miles over Hudson's Bay to the Atlantic coast.

Nest and eggs Mate for life. Three to five white eggs are laid in a shallow nest bowl of dried grasses and down feathers on the ground. Goslings remain with their parents until the following breeding season.

Voice Vocal when they are flying; emitting a rolling *Cr-r-r-nk.*

Ecological role Mainly herbivores; feed on spartina, mosses, lichens, and aquatic plants during the warmer months and on marine algae and seaweed during winter. Crustaceans, mollusks, worms, and insects are also consumed. A salt gland at the base of their bill filters salt from their blood and excretes it, which allows them to drink saltwater and feed on saltwater vegetation.

Brant geese feeding on spartina in Jamaica Bay. *DR*

Brant goose about to feed on algae, Hudson River, Riverside Park. *LD*

Canada Goose: *Branta canadensis*

Where and when to find Resident population: year-round throughout the city on its rivers, bays, sound, ponds, and lakes. Nonresidents migrate here in the spring and fly south in autumn.

What's in a name? *Branta:* burned: charcoal coloration; *canadensia*: from Canada, where it was originally identified.

Description Long black neck, white cheek patches, white throat (chin strap) and black head, brown-gray above with a whitish chest and belly and large, black, webbed feet. Males and females look alike; males are larger. Goslings are greenish yellow, round, and covered in fuzzy down feathers.

Size 26–45 inches long; wingspan: 43 to 60 inches.

Nest and eggs The female selects the site with a good view of the surrounding area and builds the nest, while the male guards it. Nests can be found on top of pilings in rivers or on the ground within 150 feet of water. The nests are bowl shaped, 1–2 feet in diameter, made from grasses, and lined with goose down from the mother's breast; 4–10 white eggs are laid over a period of days. Incubation starts when the last egg is laid so all will hatch at the same time. The parents lead the goslings to water.

Behavior Mate for life and are devoted to their goslings. The family stays together for one year, and the yearlings then separate from their parents and form groups with other yearlings. Since 2009, when a plane taking off from LaGuardia Airport landed on the Hudson River following a bird strike, thousands of Canada geese living near JFK and LaGuardia airports have been euthanized; for many birders, this is a controversial and ineffective method of dealing with the problem.

Voice Two-note *honk-haronk*. The male's voice is deeper than the female's, who has a higher-pitched honk. Canada geese are vocal, honking when they fly, when they are on the water, or when they are on the land.

Ecological role Omnivores; feed on mollusks, crustaceans, and aquatic plants and graze on land plants and grasses.

Canada geese pair flying over the Lake, Central Park.

Canada goose goslings, shore of the Hudson River near Pier I
Restaurant, Riverside Park South. Their mother led them to the
rocks to rest and preen when the tide went out.

Canada goose gosling, Riverside Park South, walking through ornamental grasses along the Hudson River.

Canada goose older goslings, Riverside Park South, near West 60th Street and the Hudson River.

Mature Canada goose, Central Park, 59th Street Pond.

Canada goose family, the Lake, Central Park.

Mute Swan: *Cygnus olor*

Where to when to find Introduced from Europe in the 1800s, they can be found during warm seasons on many of the city's ponds and lakes, such as the marsh in Inwood Hill Park in Manhattan, in the Bronx's Van Cortlandt Park Lake, on the Long Island Sound near City Island, and in large numbers in Sheepshead Bay, Brooklyn. Fall through early spring they can be seen in Jamaica Bay Wildlife Refuge.

What's in a name? *Cygnus:* swan; *olor:* swan.

Description Large, white, with orange bill, a black triangle between the eyes and the bill, and a black knob above the bill. Cygnets have gray-brown feathers and a gray-brown bill with a small black knob. Male is larger.

Size 56–62 inches long; wingspan: 7–8 feet.

Behavior Mate for life. If their mate dies, they have been known to grieve and die soon after. They have enormous, black paddle-shaped webbed feet. As they fly above you, you can hear a humming sound made by their large wings.

Nest and eggs Nesting begins in late March or April when both male and female build a large, mound-shaped nest along the shore or on an island. They use plant material and line it with feathers and down. Female incubates up to 10 eggs that turn from blue green when first laid to tan. Chicks are precocial and can walk and, within days, feed themselves, but are tended to by both parents to keep them warm and safe. They can be seen riding on their parents' back when they first hatch as their feathers are not waterproof.

Voice Mute swans are not mute. They have a call that is a soft purr, which they use with each other, with their cygnets, and with humans who feed them and whom they trust.

Ecological role Omnivores; feed mainly on aquatic plants and filamentous green algae. During winter will also feed on aquatic animals: fish, crayfish, snails, and worms.

Mute swan female on her nest, Jamaica Bay Wildlife Refuge. *JM*

Mute swan family preening, Baisley Pond Park, South Jamaica, Queens. *JM*

Mute swan cygnets, Jamaica Bay Wildlife Refuge. *DR*

Mute swan, Coney Island Creek, Brooklyn. *DG*

Mute swans with cygnets, Jamaica Bay Wildlife Refuge. *DR*

Mute swan, walking on rocks along the Central Park Lake.

Mallard Duck: *Anas platyrhynchos*

Where and when to find Our most common duck, found throughout the five boroughs of the city, in rivers, bays, the sound, ponds, and lakes.

What's in a name? *Anas*: duck; *platys*: broad or flat; *rynchos:* beak. *Mallard*: Latin for duck.

Description The male has a metallic emerald-green head, white neck ring, reddish-brown chest, grayish body, two curly tail feathers, and bright orange legs. The female has brown feathers edged in buff, a dark line from the beak across the eye toward the back of the head, and white tail feathers. Both male and female have a blue speculum edged in black and white. Ducklings are brown with yellow spots.

Size 21–28 inches long; wingspan: 30–40 inches.

Nest and eggs Males and females pair up in the autumn and engage in head bobbing and other courtship rituals throughout the winter. By March or April, the female uses grasses to build a cup-shaped nest, 7–8 inches in diameter, near the water but hidden from sight beneath a bush or tree. She lays 7–10 white eggs, which she incubates once the last egg is laid, and lines the nest with down feathers from her abdomen. When she leaves the nest to feed, she covers the eggs with these downy feathers to keep them warm and to hide them from predators. Ducklings hatch within a month and follow their mother to water.

Behavior They feed by dabbling, or "tipping up"—a method of feeding on the pond, river or lake bottom with their rumps straight up and their webbed feet paddling the surface of the water.

Voice The call of the female mallard is a loud, throaty "quack." Males have a more "reedy" voice. Ducklings have a loud piercing "peep" that can be heard for quite a distance. This helps the mother locate them if they are separated.

Ecological role Omnivores; feed on aquatic and land plants and seeds, aquatic mollusks, insects, tadpoles, and fish.

Mallard duckling walking along the shore of the Lake, Central Park.

Mallard family, the Lake, Central Park.

Mallard mother and her duckling.

Mallard couple, the Lake, Central Park.

Mallard drakes, 59th Street Pond, Central Park.

Mallard drake feeding on acorn, the Lake, Central Park.

Bufflehead Duck: *Bucephala albeola*

Where and when to find In winter on most bodies of fresh, brackish, and saltwater throughout the city.

What's in a name? *Bucephala:* buffalo headed, large headed; *albeola:* white; *bufflehead:* buffalo head.

Description A tiny duck, the smallest of the North American diving ducks; males are black above and white below with a puffy black head capped with a white crest. In sunlight, the black head has a greenish-red iridescence. The females are dark above, pale below with a white cheek patch. Males have large, white wing patches visible in flight. This white patch is less pronounced on females.

Size 13–16 inches long; wingspan: 20–24 inches.

Behavior Although tiny, the buffleheads are active ducks, scurrying across the water's surface or diving for food. Buffleheads can take off from the water like a dabbling duck, without having to run on its surface, as do most diving ducks. They are fast swimmers and can dive down to 15 feet. They are also fast fliers, flying close to 50 miles per hour. The only time they walk on land is when the female is leading her ducklings to water from their tree cavity nest.

Nest and eggs Females nest in flicker and other woodpecker holes in trees in their northern summer breeding grounds near ponds, lakes, and freshwater rivers. Males and females stay paired with the same mate for years.

Voice Females make a guttural sound while looking for nest sites and make a deep, throaty call to their young.

Ecological role Omnivores; in their summer breeding grounds, they can be found in freshwater, feeding on aquatic insects such as dragonfly and damselfly larvae and water boatman. In autumn and early winter, they consume seeds of pond plants. They spend winter diving for shrimp, snails, and other mollusks and crustaceans.

Male bufflehead, the Reservoir, Central Park.

Female bufflehead, the Reservoir, Central Park.

Northern Shoveler: *Anas clypeata*

Where and when to find A harbinger of changing seasons, this duck arrives in the city in autumn and can be seen on most wetlands, including freshwater and brackish water ponds.

What's in a name? *Anas*: duck; *clypeata*: carrying a shield.

Description A large, full-bodied duck, with a black shovel-shaped bill edged with comblike projections that filter food from the water. The male has an iridescent green head, white chest, yellow eyes, and chestnut belly and flanks, with a black back and a black bill. The female is streaked brown with brown eyes and an orange shovel-shaped bill. In flight, you can see a gray-blue shoulder patch on each wing separated by a white stripe from the emerald green speculum just below.

Size 17–20 inches long; wingspan: 30 inches.

Behavior They migrate in autumn in small flocks of 10–25 birds from the north to New York City. Pair bonding takes place during winter. By March and April, they start their return migration to the north where they reproduce. They have an unusual feeding behavior as they circle in small spinning pods, following one another, straining food from the water and muck stirred up by nearby ducks.

Nest and eggs Males accompany females as they look for a nesting site, usually in short grass near water. The female builds the nest from weeds and lines it with her down feathers. The male remains with her as she incubates 9–12 eggs for an average of 25 days. Upon hatching, she leads the precocial chicks to water where they are immediately able to swim and feed.

Voice Females make a guttural sound while looking for nest sites and make a deep, throaty call to their young.

Ecological role Omnivores; feed on plankton, insects, aquatic plants, and seeds as they swim forward with their large bills straining the water for food.

Male (L) and female (R) northern shoveler, the Lake, Central Park.

Northern shoveler groups feeding on surface of the Lake, Central Park.

Ruddy Duck: *Oxyura jamaicensis*

Where and when to find In winter can be seen on most wetlands in the city, including freshwater park ponds and brackish water ponds at Jamaica Bay Wildlife Refuge.

What's in a name? *Oxyura*: pointed tail; *jamaicensis*: found in Jamaica.

Description A small, colorful, stiff-tailed duck, during the breeding season the male has a large, sky-blue bill, broad white cheek patch, black cap and nape, and cinnamon body with white rump. Their webbed feet are far back on their body. The female is brown with a tan cheek patch and a brown stripe across it. Her bill is brown. After molting, the male's feathers are dull, and he looks like the female minus the dark stripe across the cheek.

Size 14–16 inches long; wingspan: 18 inches.

Behavior Before they dive to forage for food on the bottom, they exhale and hold their feathers against their body.

Nest and eggs Nests are typically constructed and hidden in emergent aquatic plants such as cattails or phragmites. Ruddies often weave a hanging canopy over their nests and will add more vegetation to nest if water level rises. Although they are small, they lay the largest eggs of all ducks in relation to their body size. Their eggs are creamy white, becoming tan as they age. The female incubates up to eight eggs. Only the female incubates the eggs and cares for her young. Hatchlings are precocial: covered in down feathers and able to swim and feed on their own. Until 1990, ruddies bred in Jamaica Bay Wildlife Refuge's freshwater pond, which is now brackish.

Voice Silent most of the year, males make a series of popping, "burping" noises during courtship. Females utter a series of clicks ending in a nasal *waaa* when calling their ducklings.

Ecological role Omnivores; in summer, they feed mainly on aquatic invertebrates living in the substrate, such as midge larvae, crustaceans, and zooplankton. In winter, they are more likely to feed on aquatic plants and seeds.

Male ruddy duck in his breeding plumage—sky-blue beak, cinnamon-red feather—East Pond, Jamaica Bay Wildlife Refuge. *LM*

Female ruddy ducks, the Reservoir, Central Park.

Wood Duck: *Aix sponsa*

Where and when to find In winter: lakes and ponds throughout the five boroughs. Year-round on wetlands with nesting boxes.

What's in a name? *Aix*: water bird; *sponsa*: betrothed, referring to the brilliant colors of the male's plumage, as though he were dressed for a wedding. *Wood*: nests in trees.

Description Dazzling and glorious in his breeding plumage, the male has a brilliant green iridescent head and crest, which flows down to his back. His chest is wine colored, covered with white stars, and his bill is red orange, outlined with yellow and white, with a black tip. His eyes are a brilliant orange red, and his back is dark purplish blue and iridescent. His flanks are tan, crisscrossed, and intricately patterned with brown and outlined in black and white. Female is brown with a darker head and crest and a conspicuous white eye ring. Her bill is dark and her wing feathers are iridescent coppery green in sunlight.

Size 17–20 inches long; wingspan: 28–30 inches.

Behavior Nest in tree cavities to 50 feet aboveground. Precocial hatchlings leap from their nest and follow their mother to water.

Nest and eggs Nest boxes in High Rock Park Pond in Staten Island and in Brooklyn's Prospect Park Lake; also nests in tree cavities in the New York Botanical Garden near the Bronx River. Using old nests of woodpeckers, the female pulls feathers from her breast to line the nest and incubates 10–15 creamy-white eggs. Before they pip, she makes vocalizations she will use to call them from their tree nest to the ground.

Voice The male's call is a high-pitched *jeeeee*. Female calls a soft *kuk, kuk, kuk* to her ducklings right before they hatch, when she wants them to follow her from the nest to the ground, and when they are out foraging.

Ecological role Omnivores; freshwater dabblers, feed on aquatic plants, especially duckweed and aquatic insects and tadpoles. On land, they feed on berries, seeds, nuts, and acorns. They have an expandable esophagus and feed on large acorns, which they love.

Male wood duck, full breeding plumage, the Reservoir, Central Park Reservoir.

Male and female wood ducks, 59th Street Pond, Central Park.

Male wood duck starting to show breeding plumage,
the Reservoir, Central Park.

Male wood duck in all his glory, the Reservoir, Central Park.

Male wood duck stretching his neck, which shows his
slicked-back crest, the Lake, Central Park.

Two males with one female wood duck, the Reservoir, Central Park.

Hooded Merganser: *Lophodytes cucullatus*

Where and when to find In winter and in early spring in the city's saltwater marshes and bays and in freshwater lakes.

What's in a name? *Lophodytes*: Latin for crested diver; *cucullatus*: Latin for hooded; *hooded*: high crest resembles a hood.

Description A strikingly beautiful bird. Large and flexible crests that can be raised and lowered. Male has a fan-shaped white crest bordered in black and a black head, neck, back, and tail. His breast and belly are white. He has two black stripes extending diagonally from his back, down across his white breast. His flanks are reddish brown, and his eyes are bright yellow. The female has a reddish-brown crest, dark back, and gray belly and flanks. Ducklings are precocial, covered with brown down on their back and breast. Their cheeks are tan or reddish brown, and their belly is white.

Size 16–19 inches long; wingspan: 24–26.5 inches.

Behavior As a diving duck, it is the smallest of all the mergansers. It has a visual adaptation that allows it to see well underwater. It also has a clear eyelid, or nictitating membrane, that protects its open eyes underwater as it hunts for prey.

Nest and eggs Female chooses a nest site in an abandoned woodpecker hole in a tree near water. She uses whatever material is available on the floor of the nest and adds down feathers from her belly after she starts laying her 7–13, thick-shelled, cream or yellow eggs. The male takes off as soon as she starts brooding. She must then feed herself and lead her ducklings 24 hours after they hatch, when she calls them out of the tree to the water or land.

Voice To summon ducklings, females call *croo-croo-crook*.

Ecological role Omnivores; the hooded mergansers dive for fish, crustaceans, aquatic insects, frogs, tadpoles, mollusks, and aquatic plants. On land, they and their newly hatched ducklings forage for invertebrates.

Male hooded merganser, the Reservoir, Central Park.

Female hooded merganser, the Reservoir, Central Park.

Female hooded merganser preening her feathers, the Reservoir, Central Park.

Male and female hooded mergansers, the Reservoir, Central Park.

Hooded mergansers, two resting with their bills tucked into their backs; all swimming forward, the Reservoir, Central Park.

Female hooded merganser stretching and showing her white belly and white secondary underwing covert feathers.

GULLS AND TERNS belong to the Laridae family. *Larus* is Greek for seabird. Gulls and terns have white bellies and gray or black backs. Males and females are similar, but their young are spotted or streaked. They are colonial nesters and acrobatic fliers. There are 82 species of gulls and terns worldwide, and 43 species in North America.

The three gull species covered here, great black-backed gull, herring gull, and ring-billed gull, are the most commonly seen gulls in the city. Only the ring-billed gull does not breed here. However, nonbreeding adults and juveniles live here year-round.

The common tern breeds on Governor's Island and on Long Island and spends spring and summer along the beaches of Brooklyn, Queens, and Staten Island with their young, teaching them to fish. Like gulls, these terns were almost wiped out by plume hunters in the late nineteenth century and again in the mid-twentieth century due to the insecticide DDT in the fish they fed on. In some areas along the coast, their numbers are still low, and they are listed as endangered, threatened, or of special concern in several states.

RING-BILLED
GULL

GREAT
BLACK-BACKED
GULL

Eggs shown at ½ life size.

HERRING GULL

HERRING
GULL

COMMON
TERN

Eggs shown at ½ life size.

Ring-billed Gull: *Larus delawarensis*

Where and when to find Our most common winter gull, nonbreeding adults and juveniles, can be found year-round along shorelines, but huge flocks amass fall through spring in salt and brackish waters surrounding New York City. Commonly seen inland in parking lots looking for garbage to eat.

What's in a name? *Larus*: Latin for ravenous seabird, or gull; *delawarensis*: Latin for Delaware.

Description Medium-sized gull with a white head, chest, and belly and pearl-gray wings with black tips. Two white wing spots are on the outer two black primary feathers. The yellow bill is 1.75 inches long with a black ring band around the tip. Their legs and feet are pale yellow to olive yellow.

Size Length: up to 20 inches; wingspan: 48 inches.

Behavior Acrobatic and fast, they can fly up to 44 miles per hour. These gulls can hover and appear almost stationary as they look for food. They soar on thermals and are graceful swimmers. In winter they land, run, and easily maneuver on icy ponds.

Nest and eggs Breed around the Great Lakes, in Canada, and in the Pacific Northwest. Nest constructed by both parents on the ground, open or hidden among rocks. Both parents incubate two to four olive-buff eggs with olive-brown speckles. Both parents tend to the semiprecocial young.

Voice Screeching *kree, kree;* squealing *kyow, kyow, kyow.*

Ecological role Omnivores; feed on anything they can find: fish, aquatic invertebrates, terrestrial invertebrates, and human food garbage.

Ring-billed gull profile.

Ring-billed gull first winter plumage showing streaking on head, mostly grayish back, brown wing coverts, pink bill with black tip, walking on icy Central Park Lake.

Adult, nonbreeding ring-billed gulls, showing yellow eyes, yellow bills with black ring near tip, white tail feathers, Central Park.

Adult ring-billed gull full breeding plumage with red orbital eye ring and red gape, standing on ice, the Lake, Central Park Lake.

Ring-billed gull moving into breeding plumage, showing red gape, flying over Hudson River, Riverside Park.

Adult ring-billed gull, winter plumage, flying over Central Park.

Great Black-backed Gull: *Larus marinus*

Where and when to find Year-round along the city's coast: its ocean, bays, sound, and rivers.

What's in a name? *Larus*: Latin for gull; *marinus*: Latin for sea.

Description Largest and heaviest gull in the world: white head, breast, and belly, a black back and large black wings. The large, 3-inch-long bill is pale yellow and scythe-like, with a red spot on the lower mandible toward the tip.

Size Length: to 30 inches; wingspan: to 5.5 feet.

Behavior After pair bonds are formed, male and female remain monogamous, often for years and sometimes for life. These gulls are opportunistic and clever: They will drop clams and other mollusks with hard shells on rocks and pavement to break them open.

Nest and eggs Breed in nesting colonies of wading birds, cormorants, and gulls on islands in New York Harbor, the East River, Jamaica Bay, and Long Island Sound. The pair gathers seaweed, sticks, moss, feathers, and grasses and scrapes a nest into the sand. Female lays one to five olive-buff, brown-speckled tan eggs. Parents take turns incubating for 28 days. Both parents feed the semiprecocial hatchlings. Chicks peck at red spot on adult's lower beak when they want food.

Voice The most commonly heard call is a deep, throaty, soulful *uahhh* or long drawn out, two-note *ow* repeated over and over.

Ecological role Carnivores and scavengers; great black-backed gulls feed on any animal smaller than they are, including fish, crabs, mussels, worms, other birds and eggs, and dead animals and human food garbage.

Great black-backed gull egg and hatchling, hiding in grass, island in Jamaica Bay. *JM*

Young great black-backed gull chick, running on sand, island in Jamaica Bay. *JM*

Two adult and a juvenile great black-backed gulls, feeding on algae on top of pilings, Atlantic Ocean, the Rockaways.

Adult great black-backed gull, pre-basic molt, Rockaway Beach.

Juvenile great black-backed gull, brown streaking on head and all over body, resting on top of piling, Rockaway Beach.

Adult great black-backed gull, breeding plumage, colonial nesting site, island in Jamaica Bay. *JM*

Herring Gull: *Larus argentatus*

Where and when to find Year-round along salt and brackish waters.

What's in a name? *Larus*: Latin for gull; *argentatus*: Latin for silver referring to the gray back and wings. *Herring:* gulls led fishing boats to schools of herring.

Description Large size; adult has white head, neck, tail, chest, and belly, with gray back and tops of wings. Wing tips are black with white spots. Pink legs, yellowish bill with a red spot near the tip on the lower mandible, and yellow eyes. Juveniles are spotted brownish gray, just like the young of most gulls. They go through seven molts before they develop adult plumage when they are 4 years old.

Size 26 inches long; wingspan: almost 5 feet.

Behavior Steal food from other birds. They will also drop clams and mussels from the air to crack the shells on pavement or rocks.

Nest and eggs Colonial nesters with other gulls and wading birds on small islands in Jamaica Bay, New York Harbor, and the East River. Pair chooses the nest site together, dig a "scrape," and line it with plant material, feathers, and found objects such as plastic. Parents incubate one to four pale-olive, speckled eggs. Both parents feed chicks.

Voice The long call, a trumpet-like *yeow* held for at least 3 seconds while lowering and raising their head; a short call sounds like *yelp*; warning calls: a staccato series of descending, plaintive *ha ha ha.*

Ecological role Omnivores; consume fish, aquatic invertebrates, crustaceans and mollusks, and the young of other gulls and seabirds. Beneficial as they help keep beaches clean by feeding on garbage and dead fish.

Adult herring gull, nonbreeding plumage with brown streaks on head and face, resting on Rockaway Beach.

Herring gull beach nest with two eggs, island in Jamaica Bay. *JM*

Herring gull nest with two chicks and unhatched egg, island in Jamaica Bay. *JM*

Herring gull chick pipping, colonial nesting island, Jamaica Bay. *JM*

Newly hatched herring gull chick hiding in mugwort, island in Jamaica Bay. *JM*

Adult herring gull vocalizing near nest, island in Jamaica Bay. *JM*

Common Tern: *Sterna hirundo*

Where and when to find Commonly seen from May through September along the Atlantic coast, rivers, bays, and the Long Island Sound.

What's in a Name? *Sterna*: Old English for tern; *hirundo:* Latin for swallow, for the forked tail; *common*: abundant; *tern*: Norwegian: *terna*; Danish: *tern*: for this type of bird.

Description Summer breeding plumage: light gray above, gray wings with black tips, pale gray below, black cap, orange-red legs, and orange-red bill with a black tip. Juveniles resemble adults in their winter plumage: pale gray wings with a dark carpal bar. The crown and nape are brown, and the forehead is ginger, turning white in autumn. The back is gray with ginger edges and the tail is gray and is not as swallow-like.

Size 12.2–15 inches; wingspan: 29.5–31.5 inches.

Behavior An acrobatic flier, soaring, swooping, hovering, turning quickly, diving, and plunging into the water for fish.

Nest and eggs Nest on Breezy Point beaches, Governor's Island in New York Harbor, and Joco Marsh in Jamaica Bay. Male and female choose a nest site on ground near vegetation so their chicks have some shelter from sun, wind, and predators. Female hollows out a depression, usually in the sand, and sparsely lines it with nearby vegetation and feathers. She incubates one to three eggs with great color variation: green, olive or tan, speckled with black, brown, olive brown, or gray. Hatchlings are semiprecocial, covered in down, and able to move around after a few days. They are cared for by both parents.

Voice Rattling, rolling *KEEE-aaaar, KEEE-aaaar.*

Ecological role Carnivores; eats any species of small fish that it comes upon. Also feed on invertebrates: crabs, snails, squid, and shrimp. Typically feed by diving and plunging into the water from about 9 feet up. They can also dive to the water's surface to feed or dip down into the water from the surface.

Juvenile common tern, Rockaway Beach.

Adult common tern, black cap, orange bill, Rockaway Beach.

HERONS, EGRETS, AND BITTERNS belong to the family Ardeidae; *ardea* is Latin for heron. There are 63 species worldwide. The glossy ibis is in the family Threskiornithidae. *Threskia* is Greek for religion; *ornis* is Greek for bird. The ibis was a sacred bird in ancient Egypt. There are 33 species of ibis and spoonbills worldwide.

These are large birds with long legs for wading into the water to feed. They are also colonial nesters, which offers protection from predators. They often gather in large rookeries to build their nests, lay their eggs, and raise their young. Other common characteristics are long necks and long bills. Herons and egrets have sharp bills for stabbing prey. The glossy ibis has a long, sensitive, decurved bill for probing the muck for prey.

In North America, there are 5 species of ibises (glossy, white, scarlet, white-faced, and roseate spoonbill) and 12 species of herons, egrets, and bitterns (American bittern, least bittern, cattle egret, great egret, snowy egret, reddish egret, black-crowned night-heron, great blue heron, green heron, little blue heron, tricolor heron, yellow-crowned night-heron). The following wading birds nest in New York City: black-crowned night-heron, yellow-crowned night-heron, great blue heron, little blue heron, green heron, great egret, snowy egret, and glossy ibis.

All wading birds are migratory, leaving in late summer or early autumn for their winter feeding grounds. During mild winters, you might occasionally see a great blue heron that has remained in the city.

BLACK-CROWNED NIGHT-HERON

GREAT
BLUE
HERON

GREAT
EGRET

GREEN
HERON

BLACK-
CROWNED
NIGHT-
HERON

GLOSSY
IBIS

Eggs shown at ⅓ life size.

Great Blue Heron: *Ardea herodias*

Where and when to find During spring and early fall on freshwater ponds, lakes, rivers, brackish rivers, sound, and bays. Occasionally, individuals remain in summer and through the winter. Most migrate south in autumn.
What's in a name? *Ardea:* heron; *herodias*: heron; *great blue*: large and grayish blue in color.
Description Our largest long-legged wading bird and most commonly seen heron: gray blue, white crown with wide black stripe above the eye, white around its head, neck, and chest, with some cinnamon striping on its neck. It has long dark legs, and an elongated, sharp, spear-shaped bill. In flight, it holds its head folded back in an S shape on its shoulders and trails its outstretched legs.
Size 42–52 inches long; wingspan: 6 feet.
Behavior Stands still when fishing as it waits for its prey to swim close enough for it to strike with its spearing bill. Stalking prey, this heron moves in what seems like slow motion. It is usually a solitary hunter.
Nest and eggs Rarely nests in New York City. Males gather sticks and females build the nest in a tree. The female and male take turns incubating the oval, pale bluish-green eggs. Both parents feed the hatchlings.
Voice Makes a loud *frawnk* when alarmed.
Ecological role Carnivores; feed on fish, frogs, salamanders, crabs, crayfish, grasshoppers, dragonflies, and mice.

Great blue heron hunting on ice, Upper Lobe, the Lake, Central Park.

Great blue heron flying against the backdrop of Central Park South.

Juvenile great blue heron on shore of 59th Street Pond, Central Park.

Adult great blue heron at East Pond, Jamaica Bay Wildlife Refuge. *LM*

Juvenile great blue heron on ice, Upper Lobe, the Lake, Central Park.

Juvenile great blue heron in profile, Upper Lobe, the Lake, Central Park.

Great Egret: *Ardea alba*

Where and when to find Migrates to our city in spring and stays through summer, breeding in colonies on islands around the city, and hunting for food throughout our wetlands.

What's in a name? *Ardea*: Latin for heron; *alba*: white; *egret*: from the French *aigrettes*, meaning a spray of feathers.

Description Large, white wading bird, with long black legs and black feet, a long, kinked, narrow neck and a spear-shaped yellow bill. During breeding season, long white plumes growing from the scapulars are displayed during courtship at the nest. During this period, the bill becomes a bright orange, the lores and eye rings turn lime green.

Size Almost 3 feet tall; wingspan: 51 inches.

Behavior When foraging, it drags its foot along the bottom, disturbing animals in the muck. If it sees its prey it strikes with its spear-shaped bill. They hunt on shore, in very shallow water, or in deeper water, but only up to their feathers.

Nest and eggs In nesting colonies with egrets, herons, ibises, and gulls. Males select a display area in the tree on which the nest will be built. He gathers sticks and creates a platform. Once he attracts a mate, he continues to collect sticks, which the female uses to complete the nest. Both parents incubate one to six pale greenish-blue eggs and feed the young. Courtship plumes shorten and the orange bill, and the lime-green lores and eye rings turn yellow. Sensitive to disturbance, nesting colonies are off limits to humans, except for those monitoring the nests.

Voice *Kraak* or *frawnk* calls by adults.

Ecological role Carnivores; forage mainly for fish, also invertebrates, amphibians, reptiles, small birds, and mammals. In the nineteenth century, egrets and other birds were killed for their plumes used for women's hats. By 1903, an ounce of egret plumes sold for $32, more than twice the price of gold. This practice nearly led to their extinction. The National Audubon Society was formed and successfully put pressure on Congress to pass laws to protect birds. From 1900 to about 1960, they were almost completely absent from New York. After the 1918 Migratory Bird Treaty Act was passed, their population slowly recovered, and by 1980, soon after the Clean Water Act, their population stabilized.

Great egret breeding plumage, lime-green lores and eye ring, colonial nesting island, Jamaica Bay. *JM*

Great egret hatchling and egg, colonial nesting island, Jamaica Bay. *LD*

Great egret, the Lake, Central Park.

Great egret with sunfish, the Lake, Central Park.

Great egret swallowing sunfish, the Lake, Central Park.

Great egret, the Lake, Central Park.

Green Heron: *Butorides virescens*

Where and when to find Migrates to New York City in spring and stays through summer, breeding in inland and coastal parks in trees near water. Migrates south in autumn.

What's in a name? *Butorides*: Latin for resembling a bittern; *virescens*: Latin for green.

Description Stocky and small with colorful plumage: lustrous dark-green head and back; dark blue-green wings edged in creamy white. Shaggy Mohawk-like greenish-black crests can be raised during courtship and when defending their territory. Neck and breast are chestnut with white stripes. Bill is almost 3 inches long, pointed, and sharp. Upper mandible is black; lower mandible is yellow, tinged with black. Legs and eyes are yellow.

Size 1.5 feet long; wingspan: 26 inches.

Behavior A tool-using bird, it will drop feathers or insects into the shallow water to lure fish. When the fish go for the bait, the heron grabs the fish. Can also dive into deeper water to grab fish and will roost on a branch over water and dive in to get its prey. Can slowly stalk prey or move quickly into the water.

Nest and eggs Seasonally monogamous, the pair typically bonds during migration from their winter feeding grounds in Florida and the Caribbean Islands back to New York City. Male selects the nest site and collects long, thin branches. Female arranges them in a tree near water, preferably one with branches overhanging water. A key condition for building the nest site in a woodsy area is that branches above will conceal the nest. Parents incubate two to four pale greenish-blue eggs and feed the chicks. If green herons are nesting in a colony of herons and egrets, the cover of a leafy tree is not essential.

Voice *Skeuw* and *yuk-yuk-yuk*.

Ecological role Carnivores; green herons eats earthworms, leeches, dragon-flies, damselflies, waterbugs, diving beetles, grasshoppers, crickets, spiders, crayfish, crabs, snails, frogs, toads, tadpoles, salamanders, snakes, lizards, and small rodents.

Green heron close-up, Azalea Pond, Central Park.

Green heron standing on Laupot Bridge over the Gil, the Ramble, Central Park.

Green heron nest with nestlings, Upper Lobe, the Lake, Central Park.

Black-crowned Night-Heron: *Nycticorax nycticorax*

Where and when to find Throughout the city's rivers, bays, sound, salt marshes, lakes, ponds, and New York Harbor, spring through autumn.

What's in a name? *Nycticorax*: night raven.

Description Medium-sized, stocky heron with a short, thick neck, giving it a hunched appearance, and a black cap with several long, trailing white plumes at the back of its head. It is dark gray above and white below. The eyes are ruby red, and its legs and feet are yellow. Immatures are brown with white streaks. In flight, it is a dramatic bird because of its broad wings.

Size 23–28 inches long; wingspan: 44 inches.

Behavior The black-crowned night-heron remains perfectly still as it waits for prey to swim toward it. It then thrusts its heavy bill into the water seizing fish, its main food.

Nest and eggs Breed in colonies on the small islands of New York City in large numbers. The male chooses a nest site in a tree and breaks off branches or collects them from the ground and presents them to the female, who arranges the nest with the help of the male. Eggs are greenish blue, but the color fades as the eggs age. Both male and female incubate the eggs and engage in a ceremony of vocalized greetings and feather fluffing when they relieve each other on the nest. Both parents feed the chicks.

Voice Known as the *quark* bird, because of the call it makes as it hunts, mainly at night.

Ecological role Omnivores; feed on fish, algae, and aquatic plants, frogs, toads, crayfish, blue crabs, shrimp, mollusks, and aquatic insects.

Juvenile black-crowned night-heron, Big John's Pond,
Jamaica Bay Wildlife Refuge.

Adult black-crowned night-heron in breeding plumage, showing white head
plumes, western shore, the Lake, Central Park.

Glossy Ibis: *Plegadis falcinellus*

Where and when to find April through early fall nesting on islands in New York City Harbor and feeding in salt marshes along the city's coastline, and in freshwater streams and ponds.

What's in a name? *Plegadis*: scythe or sickle, referring to the shape of its bill; *falcinellus*: liver, referring to the reddish-brown color of its feathers.

Description Medium-sized, long-legged wading bird with glossy, iridescent feathers. Head, neck, and chest are copper, and wings are bronze, green, and purple in sunlight. Adults have blue-black skin from the base of their bill to their eyes, with a dramatic edging of light-blue skin above and below the eyes. The downcurved brown bill of the glossy ibis can be almost 6 inches long.

Size About 2 feet long; wingspan: 3 feet.

Behavior Highly social, the glossy ibis flies in groups, feeds in mixed groups of other wading birds, and nests near egrets, herons, and cormorants. Colonial nesting offers protection against predators.

Nest and eggs Both parents build the nest in either coniferous or deciduous trees and shrubs and sometimes on the ground. Male brings most of the twigs and reeds to the nest. Female lines it with leaves and vegetation and incubates two to four turquoise eggs. Both parents care for the young.

Voice Nestlings beg for food using cricket-like sounds; adults croak and grunt during foraging.

Ecological role Carnivores; the glossy ibis is a tactile forager. It probes the shallow muck with its bill, feeling for aquatic invertebrates: beetles, water boatmen, dragonfly larvae, fly larvae, caddis flies, worms, mollusks (e.g., small mussels and clams), and crustaceans (e.g., crayfish, shrimp, and crabs). Also feeds on snakes, lizards, frogs, toads, and tadpoles. Originally from Africa, it migrated to South America in the nineteenth century, to Florida in the 1940s, and to New York City in the 1970s.

Glossy ibis skull, American Museum of Natural History collection. *LD*

Glossy ibis nest and eggs, colonial nesting island, Jamaica Bay. *JM*

Glossy ibis, nonbreeding plumage, feeding in Big John's
Pond, Jamaica Bay Wildlife Refuge. *DR*

Adult glossy ibis, breeding plumage, in flight over
Jamaica Bay Wildlife Refuge. *DR*

Glossy ibises on treetop, colonial nesting island, Jamaica Bay. *JM*

Glossy ibises, breeding plumage, West Pond, Jamaica Bay Wildlife Refuge. *DR*

THERE ARE APPROXIMATELY 75 species of shorebirds worldwide, mostly in the Northern Hemisphere: curlews, godwits, sandpipers, dowitchers, dunlin, knots, sanderlings, willets, yellowlegs, snipes and woodcocks, oystercatchers, phalaropes, shanks, tattlers, calidrids, and turnstones.

Most shorebirds belong to the family Scolopacidae, though oystercatchers are in the family Haematopodidae, which means blood foot, referring to the red legs of some species. *Scolopax* in Greek means woodcock. The fossil record shows that snipes, sandpipers, and woodcocks diverged into two genera 5 million to 10 million years ago: the *Gallinago* genus of snipes and the *Scolopax* genus of sandpipers and woodcocks. Woodcocks are classified as shorebirds because of their evolutionary relationship with snipes and sandpipers; however, the American woodcock has evolved to live in moist leaf litter and soil, where it uses its long bill to probe for earthworms.

Although many species of oystercatchers are striking, shorebirds typically are not colorful and have streaked brown and gray plumage above and white plumage below. They have long, sensitive straight or curved bills for probing for invertebrates.

Three commonly seen shorebirds in New York City include the American oystercatcher, the American woodcock, and the spotted sandpiper, which come to New York City to feed and breed in the spring and leave in the autumn for their winter feeding grounds in the south. Some American oystercatchers may remain as far north as central New Jersey during winter.

AMERICAN OYSTERCATCHER

AMERICAN
OYSTERCATCHER

AMERICAN
WOODCOCK

SPOTTED
SANDPIPER

Eggs shown at life size.

American Oystercatcher: *Haematopus palliatus*

Where and when to find Flying over salt and brackish waters of the city's rivers, bays, Atlantic Ocean beaches, and dunes.

What's in a name? *Haematopus*: Greek for blood foot referring to its pink legs and feet; *palliatus*: cloaked, referring to dark head and body.

Description Large shorebird with black head, dark body, white breast and flanks, and a long, straight, brilliant red-orange bill. Legs are long and pale pink. Eyes are bright yellow with a reddish-orange eye ring. The 4-inch-long bill is chisel shaped, which helps it pry open bivalves. In flight, narrow V-shaped white wing stripes and white rump are visible.

Size 19 inches long; wingspan: 32 inches.

Behavior Will often walk or run along shore rather than fly. Searches for food by wading over submerged beds of clams and mussels. When an open bivalve is found, it inserts its beak and severs the adductor muscle that holds the two shells together to consume the clam or muscle within.

Nest and eggs Nests built by both male and female on slightly elevated areas of marsh islands and beaches. Nest is a scrape. Parents use their feet to dig out a shallow depression, about 2.5 inches deep and 8 inches wide in the sand.

Voice Babies softly peep from within their eggs two days before they hatch. Young and adults make *kleep* and *hueep* calls.

Ecological role Carnivores; feed on blue mussels, ribbed mussels, soft-shelled clams, razor clams, hard clams, sandworms, and sand crabs.

American oystercatcher nest and eggs in tire detritus
on beach of colonial nesting island, Jamaica Bay. *JM*

American oystercatcher in surf, Breezy Point Beach, Queens. *DR*

American oystercatcher probing sand for food, Rockaway Beach.

American oystercatcher, wings aloft, showing
white feathers that are visible in flight.

American oystercatcher couple, female is larger than
male, Rockaway Beach, Queens.

American oystercatcher, Rockaway Beach, Queens.

American Woodcock: *Scolopax minor*

Where and when to find A harbinger of spring, can be found in park woods and meadows. Confused by city lights, they often land in unexpected places, such as Bryant Park, behind the New York Public Library, and at Rockefeller Center.

What's in a name? *Scolopax*: Latin for woodcock; *minor*: Latin for lesser. *Woodcock*: bird of the woods.

Description Black and brown leaflike patterns on their feathers, helps them camouflage in leaf litter. It has a round head and a long bill adapted for probing soil for earthworms. Their large eyes are toward the back of their heads, allowing them to see in every direction, even when they are feeding.

Size 10–12 inches long; wingspan: 16–19 inches.

Behavior When probing for worms, the woodcock rocks its body back and forth without moving its head as it slowly walks around, which may make worms move in the soil. This allows the woodcock to feel them.

During courtship rituals, which take place during dawn and dusk, a male will emit a buzzing note that sounds like *peent*. He then ascends several hundred feet in the air and spirals downward to attract a mate.

Nest and eggs Female makes her nest in a shallow depression in leaf litter in shrubby fields where she typically incubates four brownish buff/cinnamon-colored eggs with brown blotches. The male is not involved in any aspect of nesting or raising young. The female does everything. The hatchlings are precocial and leave the nest after several hours, watched over by their mother.

Voice Harsh *peents*.

Ecological role Insectivores; the flexible tip of the woodcock's bill is specialized for catching earthworms and is thought to feel worms as it probes the ground. Woodcocks digest their food quickly and an adult woodcock may eat its weight in worms every day. Earthworms provide more than 50 percent of their food and the rest of their diet includes insects, such as ants, flies, beetles, crickets, caterpillars, grasshoppers, and various larvae. They may also eat crustaceans, millipedes, centipedes, and spiders.

American woodcock walking on rocks near Azalea
Pond, Central Park.

American woodcock beak tucked under wings, resting
in garden, West End Avenue, Manhattan.

American woodcock, front view, garden, West End Avenue, Manhattan.

American woodcock under bushes, garden, West End Avenue, Manhattan.

American woodcock's back, garden, West 60s, West End Avenue.

American woodcock, side view, garden, West 60s,
West End Avenue, Manhattan.

Spotted Sandpiper: *Actitis macularius*

Where and when to find Along freshwater streams, ponds, and lakes in all five boroughs from April to May and July through September. Year-round at Jamaica Bay Wildlife Refuge, where they breed.

What's in a name? *Actitis*: shore-dweller; *macularius*: spotted; *sandpiper*: birds that chirp in the sand.

Description In breeding plumage, they have a spotted white breast, a brown back, yellow legs, and an orange bill with a dark tip. In nonbreeding plumage, they lack the spots and have an all-white breast.

Size 7.5 inches long; wingspan: 15 inches.

Behavior When foraging, they walk quickly along the shore, teetering and bobbing as they glean insects from the ground, rocks, and water's surface. This teetering walk has led to the common names of teeter-peep, teeter-bob, teeter-snipe, and tip-tail. In flight, they have a unique flight pattern over water: slow with shallow, stiff wing beats.

Nest and eggs The gender roles are reversed in this bird. Females are more aggressive and active in courtship, and the males are the primary parent, incubating the eggs and feeding the nestlings. Males have higher levels of prolactin, a hormone that increases parental care. In addition, females arrive before males on the breeding grounds, stake out territories, and attempt to attract mates. Both parents build a grass-lined nest in a depression on the ground near a body of freshwater. The female may mate with up to four males and lay eggs in several nests. Only the males incubate one to four creamy buff, speckled eggs, splotched with blackish-brown, purplish-brown, or reddish-brown spots. Hatchlings are precocial, covered with down.

Voice *Purr-weet, purr-weet*; or *weet-weet, weet-weet.*

Ecological role Carnivores; feed mainly on invertebrates: flies, grasshoppers, beetles, worms, and snails. They also feed on aquatic insect larvae and aquatic crustaceans and will capture flying insects.

Spotted sandpiper searching for food in muck,
Wagner Cove, the Lake, Central Park.

Spotted sandpiper on rocks, searching for food, the
Lower Lobe, the Lake, Central Park.

THE WILD TURKEY belongs to the family Phasianidae, named for the pheasants of the Phasis River and the ancient town of Phasis near the Black Sea. Along with the wild turkey, this family includes pheasants, peacocks, partridges, quail, grouse, and peafowl.

Most members of this bird family are Asian, European, or African. The wild turkey and the ruffed grouse are native to North America.

Characteristics of this family include colorful feathers, strong legs with leg spurs, round bodies, short, strong wings, facial wattles, and head crests. The wild turkey is the only member of this family in New York City.

WILD TURKEY

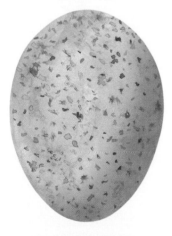

Egg shown at life size.

Wild Turkey: *Meleagris gallopavo*

Where and when to find Year-round in Van Cortlandt Park and Bronx Park. Zelda, a lone female, mainly stays in Battery Park. Hybrid turkeys are abundant in the Ocean Breeze Park neighborhood of Staten Island.

What's in a name? *Meleagros*: Greek mythological hero whose sisters were changed to guinea fowl; *gallopavo*: refers to the peacock, who also displays his tail feathers. *Turkey*: when the English first saw this American bird in the sixteenth century, they thought it was the Turkish guinea fowl.

Description Weighing up to 16 pounds with long, horizontally striped brown and white wing feathers. Dark-brown body feathers have a coppery green iridescence, and the large tail feathers are edged in chestnut brown. Featherless head is pale blue, and the face and neck are pink. Males develop scarlet-red wattles below their beaks during courtship and have bony "spurs" on the backs of their legs, which they use to fight other toms. Males have long, bristly feathered beards emerging from their breasts: the longer the beard, the older the tom. Females can have shorter beards.

Size 48 inches tall; wingspan: over 50 inches.

Behavior Highly developed senses; excellent hearing and eyesight, with enormous range of vision. Fast on land and in the air: running at 25 miles per hour and flying up to 55 miles per hour. They also swim.

Nest and eggs During mating season, males display their tail feathers in a huge and glorious fan. They strut back and forth, dragging their wings on the ground. The female digs a shallow nest in the ground hidden by shrubs or logs and lines it with grass and leaves. She incubates up to 12 eggs. The precocial poults hatch covered in fine fuzz and follow their mother to a field where they feed on insects, plants, and fruit.

Voice Tom utters a loud "gobble" over and over. Hens purr softly to their chicks while they are still in their eggs. They hatch able to discern their mother's voice.

Ecological role Omnivores; spring and summer they feed on insects, fruit, and green plants. In colder seasons, they find acorns, dried nuts, berries, mosses, buds, and ferns.

Juvenile wild turkey standing in stream near Triplets Bridge, Central Park. *DBY*

Wild turkey, Central Park.

Wild turkey resting near Diana Ross Playground, Central Park.

Wild turkey sleeping in tree, West End Avenue, Manhattan.

Wild turkey close-up, Central Park.

Wild turkey,
Central Park.

EXCEPT FOR THE OSPREY, which has been placed as the sole species in the family Pandionidae, all diurnal hawks, falcons, vultures, and eagles are members of the Accipitridae family, one of the largest of the avian families. *Accipiter* is Latin for bird of prey, specifically a hawk. The hawks in this family share the characteristic of a hooked, sharp bill for tearing prey, with a yellow cere above the top mandible. Their senses are remarkable, as they possess the "hawk" eye of legend, with the vision of some species eight times more powerful than humans. They kill with their talons and then use their beak to tear the prey apart. Females are typically larger than males. Accipiters are opportunistic feeders, taking mostly rodents and other birds as prey.

The osprey feeds mainly on fish. Its physical characteristics help to pursue fish: its outer toe is reversible, which helps latch onto slippery fish, two toes in front, two behind.

During the twentieth century, our hawks and eagles were severely affected by DDT. Fish-eating osprey and bald eagles almost vanished from North America. The DDT from the fish made their eggshells so thin that they would break when incubated. These birds have made a tremendous recovery since the banning of DDT by the federal government in 1972, 10 years after publication of Rachel Carson's book *Silent Spring*, about the deleterious effects of the pesticide DDT on humans and wildlife.

RED-TAILED HAWK

RED-TAILED   OSPREY   COOPER'S
HAWK   HAWK

Eggs shown at ½ life size.

Red-tailed Hawk: *Buteo jamaicensis*

Where and when to find Nine pairs of breeding red-tails are known to live in Manhattan, and as many as 40 pairs in the five boroughs. Pale Male, the Fifth Avenue hawk, is successfully raising young with his fifth mate after the previous four died from rat poisoning.

What's in a name? *Buteo*: a kind of hawk; *jamaicensis*: the island of Jamaica where specimens were given their scientific name.

Description Our largest hawk: dark brown above, white chest with brown streaks and chestnut-red band across the tail. Juvenile's tail is banded with brown instead of red, and they have pale eyes that mature to dark brown.

Size 17–22 inches; wingspan: 43–52 inches.

Behavior Red-tailed hawks mate for life; however, feeding on poisoned pigeons or rats and crashing into vehicles has taken the lives of many red-tails. If they die, their spouse will immediately find another mate. Red-tails are typically shy of humans but not in New York City, where these hawks nest on apartment buildings and hunt, roost, and fly near New Yorkers in parks and on city streets.

Nest and eggs Build their nests both in trees and on ledges and fire escapes of apartment buildings, churches, hotels, and skyscrapers. These nests are huge: close to 3 feet in diameter and 3 feet high, constructed with large deciduous branches and lined with fresh green sprigs in early spring. Both parents build the nest, but the female spends more time arranging the bowl, where she lays two to three brown- or red-speckled white eggs. The male provides most of the food for the family. Fledglings stay close to the parents and may even come back to the nest at night. Parents continue to provide food for the nestlings up to two months after fledging.

Voice High-pitched, hoarse, descending *keeeeeeearr*.

Ecological role Carnivores; they consume rodents, pigeons, doves, and other songbirds. When you hear crows and blue jays or see a flock of pigeons flying, look for this large hawk. Nearby birds, even tiny sparrows, will "mob" it by attacking as a group.

Red-tailed hawk feeding babies, Riverside Park.

Red-tailed hawk fledglings in London plane tree, Riverside Park.

Juvenile red-tailed hawk feeding on pigeon on Apple Bank for Savings, West 73rd Street and Broadway. Metal spikes are to keep pigeons away.

Juvenile red-tailed hawk anting, Riverside Park.

Juvenile red-tailed hawk, with young rat.

Mature red-tailed hawk bathing in puddle after a heavy rain, Riverside Park.

Osprey: *Pandion haliaetus*

Where and when to find Spring through autumn, along the coast: Brooklyn: Floyd Bennett Field and Marine Park; Bronx: the Lagoon in Pelham Bay Park; Manhattan: Hudson River; Queens: Jamaica Bay Wildlife Refuge; Staten Island: Great Kills Park.

What's in a name? *Pandion*: King of Athens; *haliaetus*: sea eagle.

Description A large raptor, dark brown above and white below. Its head is capped by a white crest with a brown cheek patch from its beak across its eye to the back of its head. Viewed from below, they have arched white wings with dark wrist patches.

Size Length: 21–26 inches; wingspan: 59–67 inches.

Behavior Perch on dead tree branches near water and fly out over water where they hover and dive. Often completely submerging themselves, they rise with fish in their talons and in midair arrange the fish so that it is head first, reducing air resistance as they fly to a perch or nest to feed the female and chicks. Fly 30–40 miles per hour but reach 80 miles per hour when diving for fish.

Nest and eggs Nesting on human-made platforms along wetlands in the city: Jamaica Bay Wildlife Refuge and Alley Pond Park, where Alley Creek flows into Little Neck Bay. Staten Island: the ponds at Mount Loretto Nature Preserve and the nest platform near River Road. Stays with mate for breeding season or, in some cases, for life. Male carries sticks and female arranges huge nests reaching 12 feet deep and 6 feet in diameter on top of trees or more shallow nests on human-made platforms near water. Female lays two to four mottled cream and tan eggs speckled with cinnamon brown. Male feeds female as she incubates and feeds entire family for up to two months, carrying three to six fish a day to the nest.

Voice High-pitched *cheep-cheep-cheep*.

Ecological role Carnivores; herring, bluefish, eel, flounder, menhaden, perch, shad, and bass. During the middle of the twentieth century, osprey were almost driven to extinction by the pesticide DDT, which collected in the tissue of the fish they ate: the DDT in their systems caused a problem with calcium production creating eggs with very thin shells that broke easily during incubation. Since the banning of DDT in 1972 and the Endangered Species Act of 1973, ospreys have made a comeback.

Juvenile osprey, Jamaica Bay Wildlife Refuge. *DR*

Osprey nest, Jamaica Bay Wildlife Refuge,
John F. Kennedy International Airport behind. *DR*

Osprey and chicks. *MP*

Juvenile osprey triplets still in nest, Jamaica Bay Wildlife Refuge. *DR*

Juvenile osprey, Jamaica Bay Wildlifc Refuge. DR

Adult osprey flying with fish to feed its family in the nest,
Jamaica Bay Wildlife Refuge. DR

Cooper's Hawk: *Accipiter cooperii*

Where and when to find In parks and nature preserves in all five boroughs: August through April.

What's in a name? *Accipiter*: Latin for bird of prey; *cooperii*: for William C. Cooper, a zoologist and founder of the New York Lyceum of Natural History in 1817, renamed the New York Academy of Sciences. Members went on to create New York University in 1831 and the American Museum of Natural History in 1868. Both Charles Darwin and John James Audubon were members of the New York Lyceum.

Description A woodland hawk that skillfully and quickly flies in between trees on broad, rounded wings and a long, rounded tail. Adults are steely blue above, with reddish bars across their breast and belly, and wide, dark bands on their tails. Juveniles have brown feathers edged in white above with brown streaks down their breast. Second-year individuals are brown turning gray above. Closely resembling the sharp-shinned hawk, Cooper's hawks are much larger.

Size 14–19 inches long; wingspan: 28–34 inches.

Behavior Their flight pattern is flap-flap-glide. During an attack, it flies fast and low to the ground, suddenly appearing to surprise its prey.

Nest and eggs Nests in the Bronx and Staten Island. Male builds large nest 20 feet to 60 feet high in pine trees using large twigs to create a platform. Nest is lined with bark. Female incubates four to six pale white eggs tinged with blue. Male feeds female, who tears prey up for hatchlings. Nestlings are semialtricial and covered in white down.

Voice A steady stream of *cak-cak-cak-cak-cak-cak-cak-cak*, lasting up to 5 seconds.

Ecological role Carnivores; they feed mainly on birds, particularly medium-sized birds: European starlings, mourning doves, rock doves, American robins, blue jays, and northern flickers. They will rob nests and eat small mammals, such as chipmunks, mice, squirrels, and bats.

Juvenile Cooper's hawk feeding on male cardinal.

Juvenile Cooper's hawk walking on rock, Central Park.

Adult Cooper's hawk feeding on pigeon, Wakefield Park, the Bronx. DG

CARACARAS AND FALCONS belong to the family Falconidae, which includes 60 species of these diurnal birds of prey worldwide. They are found in diverse habitats, including wetlands, mountains, forests, deserts, suburbs, and cities. In North America, we have the gyrfalcon—the largest falcon; the peregrine; the merlin; and the American kestrel, our smallest falcon. In New York City, the peregrine and the American kestrel are extremely common, the merlin, less so. All falcons have tapered wings, enabling them to fly at extremely high speeds. The peregrine falcon is the fastest animal on Earth.

All falconids kill with their beaks, not their talons. Most species are monogamous, with the female doing the incubating, brooding, and feeding of young. The male hunts and brings the food to the nest.

The falcons that live in New York City have adapted well. In the wild, peregrines normally nest on rocky cliffs and bluffs. As a result, they have made their nest on stone buildings and bridge ledges throughout the five boroughs. American kestrels are cavity nesters using abandoned woodpecker holes as nests. In the city, they use hollow metal building cornices for their nests. There are more peregrine falcons and American kestrels in New York City than in any other urban site in the world.

PEREGRINE FALCON

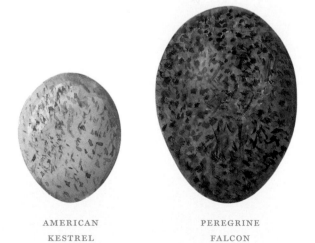

AMERICAN
KESTREL

PEREGRINE
FALCON

Eggs shown at life size.

American Kestrel: *Falco sparverius*

Where and when to find The most abundant raptor in the city; year-round in the five boroughs, on top of church spires, building antennas, or water towers; any high point where they can look for prey.

What's in a name? *Falco*: Latin for sickle, referring to the talons' shape; *sparverius*: Latin for sparrows, one of the birds they feed on. *Kestrel*: Latin for chattering bird.

Description Our smallest falcon, kestrels are dimorphic: the male has blue-gray wings, rufous back with horizontal black bands, rufous rump, and a long rufous tail, ending in a wide horizontal black band with white tips. The female has brown wings, with horizontal black or grayish bars. American kestrels have a grayish-blue crown, with a rufous patch, and two vertical black stripes on their faces with a white patch in between. Breeding males have an orange cere. Females are larger than males.

Size 10 inches long; wingspan: 2 feet.

Behavior Kestrels hover over their prey and swoop down to catch it with their talons. They fly with a rowing movement of their wings.

Nest and eggs Cavity nesters, these small falcons often find openings in cornices on old city buildings. In parkland, they use woodpecker holes. They gather no nesting material. The female makes a scrape in whatever material is found on the nest floor and lays four to five eggs varying in color from white to cream to yellow to pale pink, sprinkled with fine, reddish-brown spots. Female incubates eggs. Both parents feed young, but male does the hunting.

Voice High-pitched *killy-killy-killy.*

Ecological role Carnivores; mainly insects and mice.

Adult male American kestrel. *JO*

Adult male American kestrel feeds on its sparrow lunch on a windowsill, while Mr. Darcy, the cat, looks on, Washington Heights. *SS*

Peregrine Falcon: *Falco peregrinus*

Where and when to find In the wild, these falcons nest on rock ledges high above their hunting grounds. The city, with its bridges, skyscrapers, and steeples, offers nesting and hunting sites, making New York the city with the highest density of peregrines in the world.

What's in a name? *Falco*: falcate or hooked shape of talons; *peregrinus*: "wandering" as peregrines can be found worldwide.

Description A large, powerful raptor, with sharply pointed wings, narrow tail, white chest, and an abdomen heavily spotted and barred with dark gray. Head with black crown; face with black cheek patches, or "sideburns."

Size 14–18 inches long; wingspan: 37–46 inches.

Behavior During a stoop, peregrines can fly at a speed over 200 miles per hour when diving for prey. They are the fastest animals on Earth.

Nest and eggs Some nest sites: the towers of the Marine Parkway Bridge in Brooklyn; Riverside Church, the MetLife Building, and 55 Water Street in Manhattan; the Goethals Bridge in Staten Island. Nest is called a *scrape*, a shallow depression on a rocky ledge, deep enough so that the three to four eggs covered in rich browns, reds, or purple splotches don't roll out. Parents feed new hatchlings only the muscles of birds, without bones. After a few weeks, the chicks are fed on whatever parents are eating.

Voice Harsh *kak kak kak kak kak* over and over.

Ecological role Peregrines have made a remarkable recovery since DDT was banned and the Endangered Species Act was passed, which helped bring them back from near extinction. Powerful carnivores, feeding on pigeons, mourning doves, and a variety of songbirds. New York City peregrines feed on more than 75 species of birds.

Peregrine falcon fledgling, Gil Hodges Bridge, Brooklyn. *DR*

Adult peregrine falcon, Gil Hodges Bridge, Brooklyn. *DR*

Juvenile peregrine falcon, Brooklyn. *DR*

Peregrine falcon flying. *DR*

Peregrine falcon rescued by Bobby and Cathy Horvath outside Payson Park House, Inwood Hill Park.

Riverside Church male peregrine falcon flying over Morningside Heights. *JO*

MOST OWLS belong to the family Strigidae. The Latin meaning of the word *strix* is owl. Two hundred species of owls in this family share physical characteristics. They are nocturnal hunters but will also hunt during the day. They have strong, hooked bills used for tearing prey, round facial discs for collecting sound, large eyes for collecting light, fairly short legs, strong feet, and sharp talons. Most species have feathered ear tufts. They live on every landmass worldwide from the Arctic to the tropics, from sea level to high in the mountains.

The barn owl belongs to a separate owl family: the Tytonidae. *Tyto* is Greek for night owl. There are 17 species of barn owls, and they all have large heads, heart-shaped facial discs for collecting sound, long legs and strong feet. Barn owls do not have feathered ear tufts. They also live worldwide, but not in Canada, northern Asia, the tundra, Iceland, or the Arctic.

North American owls include the screech, spotted, crested, pygmy, barred, spectacled, elf, saw-whet, long-eared, short-eared, great horned, snowy, and boreal. The most commonly seen owls in New York City are the great horned owl, the barred owl, the screech-owl, and the tiny saw-whet owl. Barn owls can be seen in Jamaica Bay Wildlife Refuge, where nesting boxes have been built for them. Occasionally, snowy owls show up during winter when food is scarce in the Arctic tundra. During the winter of 2013–14, snowy owls were observed in great numbers along the coast in Brooklyn, the Bronx, Queens, and Staten Island.

NORTHERN SAW-WHET OWL

EASTERN SCREECH
OWL

GREAT
HORNED
OWL

BARRED
OWL

NORTHERN
SAW-WHET
OWL

Eggs shown at ⅔ life size.

Eastern Screech-Owl: *Megascops asio*

Where and when to find Year-round; wooded parks, Alley Pond Park, Queens; Pelham Bay Park, the Bronx; Kingfisher Park, Staten Island.

What's in a name? *Megascops*: large eared; *asio*: horned owl; *screech*: one of the sounds they make.

Description A small, starling-sized owl, with large tufts that resemble pointed "ears," a large head, seemingly no neck, rounded wings, large yellow eyes, feathered yellow feet, and a short tail. There are two color morphs: gray and reddish brown, both with patterns of stripes and spots that give this owl amazing camouflage against a tree.

Size 8.5 inches long; wingspan: 20 inches.

Behavior Nocturnal and vocal at night, these little owls are heard more often than seen. They are often sedentary hunters, waiting on their branch for prey to pass by. Then they pounce. When food is abundant, they store uneaten prey in tree holes for days.

Nest and eggs Nesting in the Bronx and Staten Island. These owls generally mate for life and are cavity nesters. They use abandoned woodpecker holes, where the female nests on whatever debris is on the cavity floor. She incubates two to eight white eggs and is fed by her mate. The male hunts for prey, and the female rips it apart and feeds it to the nestlings.

Voice Both male and female sing a captivating and unearthly trill ending in a descending whinny.

Ecological role Carnivores; they feed on almost any animal they encounter: thrushes, starlings, waxwings, finches, flycatchers, jays, doves, woodpeckers, swallows, bats, rabbits, mice, squirrels, rats, and moles; in summer, primarily invertebrates: worms, insects, crayfish, and vertebrates, such as toads, salamanders, and frogs.

Eastern screech-owl fledglings, Central Park.

Adult eastern screech-owl, gray morph, Central Park.

Great Horned Owl: *Bubo virginianus*

Where and when to find Throughout the year in our large parks with old growth trees such as Pelham Bay Park, Inwood Hill Park, and the New York Botanical Garden.

What's in a name? *Bubo*: Latin for owl; *virginianus*: of Virginia, where it was first identified; *great horned*: large owl with ear tufts.

Description Large, heavy, barrel-shaped owl, with cinnamon-red facial disc, bright yellow eyes and a large white throat patch. Prominent feather tufts on top of head, which can be lowered. Body is mottled brown above and light brown below. Feathers completely cover legs and feet, which muffles the sound of flight as they strike prey with their talons.

Size 17–25 inches long; wingspan: up to 5 feet.

Behavior Pair may stay in the nesting territory throughout the year. Loud hooting in defense of their territory begins in early winter before egg laying and again in the autumn when fledglings leave nest. Fierce predators, they grip their prey with such strength that it takes a force of almost 30 pounds to open their talons, which they use to sever their prey's spine.

Nest and eggs Mate for life and nest in the winter on South Brother Island, Pelham Bay Park, New York Botanical Garden, Inwood Hill Park, and Staten Island. They use abandoned nests of red-tailed hawks and other birds and squirrels. There is an elaborate courtship of bowing and bill rubbing. Female incubates one to four round, white eggs. Male brings food to her and to the hatchlings.

Voice Booming *ho-ho-hoo hoo hoo.*

Ecological role Carnivores; 90% of their diet comes from mammals: they are the only predator of skunks. They also feed on rats, chipmunks, woodchucks, porcupines, house cats, and some birds, including ducks, crows, hawks, and starlings. Although mainly nocturnal, they will hunt during the day.

Great horned owl fledgling, New York Botanical Garden, the Bronx. *LM*

Adult great horned owl, Central Park.

Barred Owl: *Strixa varia*

Where and when to find Wooded parks in late fall and winter: New York Botanical Garden and Pelham Bay Park in the Bronx; Central Park in Manhattan.

What's in a name? *Strix*: Latin for owl; *varia*: variegated feather patterns.

Description: Large with a round head, no ear tufts, and soulful, dark eyes. Body: varied patterns of brown and white overall. The back and chest have horizontal brown bars on a white background. Belly: vertical brown bars on white. Wings and tail are barred brown and white.

Size 16.9–19.7 inches long; wingspan: 39–43.3 inches.

Behavior They have been known to jump into shallow water to catch fish. Nocturnal, but they occasionally hunt during the day. They swallow small prey whole and will rip the head off larger prey, devouring that first and then eating the rest.

Nest and eggs Have nested in Fort Tilden, Queens. Mate for life and use natural tree cavities and abandoned hawk or crow nests. Female incubates two to three round, white eggs and is fed by her mate. Hatchlings are semialtricial. Both parents care for their young.

Voice: *Hoo, hoo, too-HOO; hoo, hoo, too-HOO, ooo,* or *who cooks for-you— who cooks for you-all,* with the last note dropping off.

Ecological role: Carnivores; feed on mammals: voles, squirrels, chipmunks, mice, and rabbits; birds: woodpeckers, blue jays, and pigeons; amphibians: frogs, toads, and salamanders; reptiles: turtles and snakes; invertebrates: beetles, grasshoppers, and crickets.

Barred owl sleeping, Central Park.

Barred owl, Bronx Zoo.

Barred owl, preening foot, Central Park.

Barred owl in wild black cherry tree, Central Park.

Barred owl sleeping in white pine tree, Central Park.

Barred owl, eyes open, Central Park. There is a need for quiet and respect when near owls.

Northern Saw-whet Owl: *Aegolius acadicus*

Where and when to find In winter: Pelham Bay Park in the Bronx; Central Park in Manhattan; Prospect Park, Brooklyn.

What's in a name? *Aegolius*: Greek for owl; *acadicus*: Latin for "of Acadia"; early European settlers discovered this owl in the North American colony Acadia, which is now Nova Scotia. *Saw-whet*: alarm call sounds like a saw being sharpened on a whetstone.

Description This small owl has a rounded head, no ear tufts, a white face surrounded by brown and white feathers. Its breast and belly are white streaked with reddish brown. They have large yellow-amber eyes.

Size 7–8 inches long; wingspan: 16–19 inches.

Behavior If you suddenly encounter a northern saw-whet owl, it will sit very still, a defensive adaptation against predators that attack prey in motion.

Nest and eggs Female incubates 4–10 white eggs in an abandoned woodpecker tree cavity. Male feeds female while she is brooding. He continues to bring food to the nest and passes it to her. She tears it into small pieces to feed her young.

Voice Many vocalizations, among them, their song: a rapid series of whistled *toot-toot-toot-toot-toot*; alarm call: high-pitched *ksew-ksew-ksew*, repeated many times.

Ecological role Carnivores; they feed mainly on mice and occasionally on birds and insects.

Northern saw-whet owl sleeping, Shakespeare Garden, Central Park.

Northern saw-whet owl, Shakespeare Garden, Central Park.

Northern saw-whet owl named Cricket, rescued by Bobby and Cathy Horvath, at Inwood Hill Park educational event.

Northern saw-whet owl, Central Park.

Bobby Horvath holding rehabilitated northern saw-whet owl before release. *JO*

Release of northern saw-whet owl from Bobby Horvath's hands, Inwood Hill Park. *JO*

IN THE AVIAN family Trochilidae, there are 328 species of hummingbirds worldwide. These are typically tiny birds whose long bills fit into the nectaries of flowers. Bill shape and length indicates the shape of the flower the hummingbird visits. Although hummingbirds feed primarily on flower nectar, they will supplement their diet with small invertebrates. They have unique evolutionary adaptations, such as long, thin bills; extendable tongues; and the ability to hover, all characteristics that help them consume nectar.

Hummingbirds are dimorphic: The male is typically more colorful than the female, and the female is slightly larger. They have physical adaptations that allow them to hover and can fly both forward and backward. The humming sound they make that gives them their common name is caused by modified outer primary feathers. They are fast and can fly at almost 30 miles per hour. Their wingbeat can reach 80 beats per second. Their heart can beat up to six times per second while at rest and up to 16 beats per second when active.

Hummingbirds are solitary. After mating, the female builds the nest, broods, feeds the young, and protects her nest site. Males are pugilistic and will patrol their food source and aggressively defend it against intruders, as anyone with a hummingbird nectar feeder will tell you.

Tropical hummingbirds do not migrate, but the 17 species of hummingbirds of North America can migrate up to 3,000 miles round trip to and from their southern wintering grounds, including the one species that lives east of the Mississippi River: the ruby-throated hummingbird. This tiny jewel of a bird is commonly seen in the parks and backyard gardens in New York City. In autumn, it migrates to southern Mexico, to Central America, as far as Costa Rica, and to the West Indies.

RUBY-THROATED HUMMINGBIRD

Egg shown at life size.

Ruby-throated Hummingbird: *Archilochus colubris*

Where and when to find These miniature birds can be found throughout the spring and fall migration months in city parks and community gardens, feeding on the nectar of flowers. They linger here in the fall before they start their migration south.

What's in a name? Named for the ancient Greek poet Archilochus; *colibri*: French and Spanish word for hummingbird.

Description Both male and female ruby-throated hummingbirds are metallic green above and gray below. The male has a gorget (pronounced gor-jet): iridescent throat feathers that appear brilliant red in the sun and black in the shade. Females have a white throat. Male's tail is slightly forked.

Size 3–3.75 inches long; wingspan: 4–4.75 inches.

Behavior Travel hundreds, sometimes thousands, of miles during their fall and spring migrations. In spring, they time their migration north to that of the yellow-bellied sapsucker, because it drills sap holes in tree trunks. The ruby-throats drink from the sap holes until the flowers they depend on bloom. The wild columbine times its flowering to coincide with the arrival of the ruby-throat. The long, red petal spurs provide nectar, and the hummingbird spreads the pollen. Ruby-throated hummingbirds can beat their wings more than 70 times per second and more than 200 times per second during courtship displays. The hum of their wings gives them their common name.

Nest and eggs The only hummingbirds that live east of the Mississippi River, the female constructs her nest on the branch of a tree. Usually impossible to see, the elfin nest looks like a pale knob and is made of spider silk and thistle or dandelion down, camouflaged with pale-green lichen. The female incubates two white eggs the size of jelly beans, and the spider silk expands as the babies grow. Only the female cares for the young. Though hummingbird nests are rarely found in New York City, one female built her jewel of a nest in Central Park in the spring of 2014. Unfortunately, her tree was near an oriole nest and the oriole consumed her eggs.

Voice A tiny squeak that they utter repeatedly when other ruby-throats approach a flower or the nectar feeder they are drinking from.

Ecological role Pollinators; feed on the nectar of flowers and carry pollen from flower to flower. They also consume insects and spiders, which they feed to their babies.

Male ruby-throated hummingbird, gorget aflame, Central Park.

Female ruby-throated hummingbird taking nectar from jewelweed flower, while hovering, Strawberry Fields,
Central Park. Jewelweed nectar is among their favorite.

Female ruby-throated hummingbird at cardinal flowers, Lower Lobe, the Lake, Central Park. Note the white tips of her tail feathers. Males' tail feathers are brown.

Ruby-throated hummingbird sits in her nest made of spider webs, lichen, and dandelion down, on branch of sweetgum tree, Central Park. Sadly, the eggs were eaten by a Baltimore oriole from a nearby nest.

Juvenile male ruby-throated hummingbird. He, like the female, has white-tipped tail feathers. He also has a tiny red gorget, which will cover more of his throat as he matures during his first winter.

Female ruby-throated hummingbird cleaning her bill.

PARROTS belong to the family Psittacidae. *Psittakos* is Greek for parrot. The old world, or Afrotropical, parrots of the subfamily Psittacinae consist of 11 species, including the African grey and the Senegal parrots. The new world, or Neotropical, parrots of the subfamily Arinae consist of 148 species, including macaws, conures, Amazon parrots, and the monk parakeet.

Afrotropical species live in southeast Asia, Africa, and Australia. Neotropical species live naturally in Mexico, the Caribbean, and South America. Parrots are intelligent birds, capable of imitating human speech. Because of habitat loss and exploitation by the pet trade, parrots are more endangered than any other type of bird. The numbers of parrots living in the wild have been greatly reduced worldwide; however, associations such as Defenders of Wildlife, the Humane Society International, and the Avian Welfare Coalition are working to help parrots survive in the wild. The monk parakeet, native to South America, thrives in New York City. The enormous colonial nests they build are warm in the winter. Mainly herbivores, they feed on buds, flowers, fruit, and berries.

MONK PARAKEET

Egg shown at life size.

Monk Parakeet: *Myiopsitta monachus*

Where and when to find The Bronx: Pelham Bay Park; Brooklyn: Green-Wood Cemetery, Brooklyn College, Marine Park; Manhattan: West Side; Queens: Howard Beach; throughout Staten Island.

What's in a name? *Myia*: a fly; *psitta*: parrot; *monk*: like a monk's hood—gray forehead and neck.

Description Bright green above with a long, pointed, green and blue tail. Its upper belly is lemon to olive yellow; lower belly is bright green. Its wings are blue. The very top of its head, neck, and breast are gray. Its large bill is flesh colored.

Size 11–12 inches long; wingspan: 17–18 inches.

Behavior Native to South America, they were brought in for the pet trade in the early 1970s, and because they are such social birds, it is believed that as they escaped their homes, they began to congregate and reproduce. Monk parakeets successfully live and breed in New York City because they construct and roost in enormous colonial nests. These nests, often more than 6 feet long and 3–4 feet wide, keep them warm in the subfreezing winters and cool in the summer. Other than eating and preening, all colonial activity is geared toward caring for the nest. All birds, young and old, contribute to building, repairing, and cleaning their nests.

Nest and eggs Pairs mate for life and construct separate nesting chambers with their own entrance. Within this chamber, the female incubates five to eight smooth, white eggs. Male brings food to her during incubation, and both parents feed nestlings.

Voice These highly social birds have a wide array of calls: at least 11: threat, alarm, greeting, flight, contact, preening, isolation/fear, chatter, distress, food-begging, and feeding.

Ecological role Herbivores; feed mainly on seeds, fruit, berries, nuts, buds, flowers, and occasionally on invertebrates. Adaptable to a variety of food sources.

Monk parakeet pair in front of their colonial nest,
Green-Wood Cemetery Gate, Brooklyn.

Monk parakeet feeding on mulberries, Lincoln Towers garden, West End
Avenue. This monk spent two days consuming mulberries

Monk parakeet colonial nests in the famous gates of Green-Wood Cemetery. These Gothic Revival structures, made of Belleville brownstone, were built in 1861. Monk parakeets have nested within them for years.

Monk parakeet, garden, West End Avenue.

Monk parakeets fly in small flocks with level wings. These are Green-Wood Cemetery birds.

Lone monk parakeet bathes in puddle with European starlings, Green-Wood Cemetery.

THE COLUMBIDAE family includes pigeons and doves. *Columba* is Latin for pigeon or dove. *Pigeon* in Old French meant young dove; in Latin, *pipio* means to chirp. Dove, in Anglo Saxon, is *dufan* for diving flight. Worldwide there are 310 species of pigeons and doves, including the rock pigeon and mourning dove so common in New York City. The larger species in the Columbidae family are often called pigeons and the smaller ones, doves, but this is not based on any scientific taxonomy and is interchangeable in some places.

Pigeons and doves have plump bodies and short necks. The family is divided into seedeaters and fruit eaters. Seedeaters, like our rock pigeons and mourning doves, typically have tan, gray, or brown feathers, and the feathers of fruit eaters are usually colorful greens and oranges, such as the rose-crowned fruit dove of Asia and Oceania. Pigeons and doves can have iridescent feathers around the face, neck, back, wings, and breast.

Almost all members of this family are monogamous and raise their chicks together. Both males and females produce crop milk, which consists of water, protein, fat, and minerals. Babies grow quickly on this nutritional diet and after a few days are fed seeds and fruit.

Pigeons and doves have extraordinary navigational skills and have served humans as messengers for thousands of years.

MOURNING DOVE

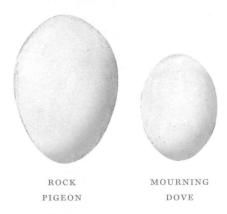

ROCK
PIGEON

MOURNING
DOVE

Eggs shown at life size.

Rock Pigeon: *Columbu livia*

Where and when to find Everywhere, in every season: streets, parks, and buildings.

What's in a name? *Columba*: Latin for dove; *livia*: Latin for lead colored; rock: pigeons nest on stone ledges.

Description Color variations: bluish gray, charcoal gray, reddish gray, and cinnamon red; iridescent neck feathers of the hackles shine reddish purple, green, and pink in sunlight.

Size 12.5 inches long; wingspan: 28 inches.

Behavior Mate for life; during courtship dances, male puffs up his hackles, bows to his mate, and spins around. After mating, the male often takes flight, clapping his wings. Acrobatic flyers, they can reach speeds of more than 50 miles per hour. Pigeons were the first domesticated birds thousands of years ago because of their intelligence, incredible eyesight, and homing skills. Early European settlers brought them to North America in the seventeenth century. As carrier pigeons, they have saved thousands of lives during wars.

Nest and eggs Devoted parents; the male chooses a site on a flat stone ledge with an overhang. The male incubates two white eggs midmorning through late afternoon and the female late afternoon to midmorning. Both parents feed their babies "pigeon milk," produced in their crops the first week of the nestlings' lives, before they can digest solid food. Crop milk is rich in protein, fat, and calcium. After a week, the parents include seeds with the crop milk, and eventually, they only feed their nestlings regurgitated seeds.

Voice *Coo roo-c'too-coo.*

Ecological role Omnivores; they help clean up food litter on our streets and feed on berries and weed seeds.

Rock pigeons mate for life. Here they are involved in alloprcening.

Squabs in the nest with bulbous beaks and
sparsely covered in stringy yellow down.

Parent feeding juvenile rock pigeon regurgitated seeds.

Gorgeous colors of the mature rock pigeon, amber-red eyes,
iridescent feathers, white cere.

Red and white morph rock pigeon after a rain. Bathing in rain, the birds erect their feathers, spread their tails, and raise their wings so that the water reaches every part.

Pied morph rock pigeon drying after a rain.

Mourning Dove: *Zenaida macroura*

Where and when to find Parks, backyards, year-round.

What's in a name? *Zenaida:* named for Zenaida Bonaparte, wife of nineteenth-century French ornithologist Charles Bonaparte; *macroura:* Greek for long tail.

Description Small, round head, plump body, long tail with black border and white tips, tawny head and breast, blue featherless skin around eyes, buffy wings with black spots, short red legs and feet, pink iridescence on neck of males.

Size 12 inches long; wingspan: 18 inches.

Behavior May pair for life, but most stay with the same mate throughout a breeding season. During fall and winter, many doves can be seen feeding together. By spring, they choose their mate. A fast flyer, the mourning dove has clocked at 55 miles per hour.

Nest and eggs Loosely arranged sticks on building ledge. Sticks gathered by male, but arranged by female, who lays two white eggs. Incubation is about two weeks. Male sits on eggs during the day, female sits on eggs at night. Both parents feed baby birds crop milk until they are old enough to digest seeds.

Voice *Coo ahhhh coo coo coo*, with a lift up at the *ahhhh*.

Ecological role Herbivores; weed seeds. Can be seen year-round combing the grassy areas of parks and gardens and showing up on the ground below bird feeders. Like their relatives, the rock pigeon, they drink using their beaks like straws, sucking up water, rather than tilting their heads back to swallow as most birds do. Mourning doves are commonly eaten by birds of prey such as red-tailed hawks.

Male mourning dove is more colorful than female, but both have blue skin around their eyes.

Female in nest built atop fuse box feeding nestling.

Female feeding two nestlings simultaneously.

Young mourning dove forming adult plumage, skin around
eyes turning from gray to blue.

Parent with nestling in tree nest, Lincoln Towers garden,
West End Avenue.

Adult mourning dove sleeping in leaf litter, Central Park.

THERE ARE 200 SPECIES IN THE FAMILY PICIDAE, which includes woodpeckers, sapsuckers, flickers, wrynecks, and piculets. They are found in most places around the world except in Madagascar, New Zealand, Australia, and the poles. Twenty-three species of flickers, sapsuckers, and woodpeckers can be found in North America.

Shared physical characteristics include a strong, straight, chisel-shaped bill; a thick, protective spongy skull so that they can chisel wood and drum without concussing; short strong legs and zygodactyl toes for climbing tree trunks; and stiff tail feathers, or retrices, which support them as they move vertically up trees. They have long, sticky tongues, which they use to probe and grab invertebrates hiding in bark crevices.

Woodpeckers drum against almost any hard surface to announce their territories and communicate with a mate. Both parents excavate their nest cavity, and both parents care for their young. Whether excavating or drumming, woodpeckers can peck up to 20 times per second.

New York City has five species in the Picidae family: the downy woodpecker, the hairy woodpecker, the red-bellied woodpecker, the yellow-bellied sapsucker, and the northern flicker. Woodpeckers, which feed on nuts, seeds, and invertebrates, can be seen year-round in our parks and wherever there are trees. Sapsuckers and flickers migrate south to their winter feeding grounds.

YELLOW-BELLIED SAPSUCKER

RED-BELLIED
WOODPECKER

YELLOW-
BELLIED
SAPSUCKER

DOWNY
WOODPECKER

HAIRY
WOODPECKER

NORTHERN
FLICKER

Eggs shown at life size.

Red-bellied Woodpecker: *Melanerpes carolinus*

Where and when to find Lives throughout the five boroughs in the large and smaller parks, year-round.

What's in a name? *Melanerpes*: black-headed creeper; *carolinus*: where this bird was first illustrated; *red-bellied*: pale red spot on its lower belly, which is difficult to see.

Description A large woodpecker with a pale face, chest, and abdomen. The male has a brilliant orange-red crown that goes from his bill to his nape. The female has a pale-red nape and a red spot over her bill. Both male and female have the black and white "zebra-back" pattern of many woodpeckers. Their stiff retrices, used to prop them against the trunk, have a similar black and white pattern. Their toes are zygodactyl.

Size 9–10.5 inches long; wingspan: 15–18 inches.

Behavior Live and nest throughout the city. When choosing a nesting site, the male will drum three taps per second. If the female accepts the site, she will engage in mutual drumming. No tapping means that she rejects the nest site.

Nest and eggs Once the nesting site is accepted, they use their chisel-like bills to excavate cavities in dead trees. They also nest in soft-wood city trees, such as lindens. The cavities are up to 12 inches deep. Four glossy white eggs are laid directly on wood chips. Females stay with the young during the day, and the males are with them throughout the night. Nestlings are fed insects and fruit.

Voice Rolling, bubbling, rising *quir?*

Ecological role Omnivores; feed on invertebrates, berries, fruit, nuts, and seeds. They use their powerful bills to hammer on tree trunks and branches for wood-boring insects. During winter, these woodpeckers will visit bird feeders for seeds and suet.

Male red-bellied woodpecker preening. It is rare to see the red patch on the belly.

Juvenile red-bellied woodpecker.

Yellow-bellied Sapsucker: *Sphryapicus varius*

Where and when to find In parks throughout the five boroughs, during spring and fall migration: on trunks of street trees, and in wooded areas of city parks and backyards.

What's in a name? *Sphrya*: *Greek* for hammer; *picus:* Latin for woodpecker; *varius*: Latin for variegated coloring of white bars on black back and wings; *sapsucker*: drinks tree sap.

Description Black head with white lines down the side and a red forehead and crown; a black back spotted with white, a pale-yellow breast and upper belly, and a conspicuous white wing patch. Males have red throat patch; females have white.

Size 7.1–8.7 inches long; wingspan: 16 inches.

Behavior Male and female drill rows of squarish sap holes in late winter or spring and tap xylem sap, which carries water and nutrients from roots up to the newly emerging tree leaves. In summer, they feed on phloem sap that carries nutrients produced in the leaves down to all parts of the tree. Sapsucker saliva may contain an anticoagulant preventing sap from clogging up. They often choose trees damaged by insects, disease, or lightning: the sap of sick trees contains higher levels of proteins.

Nest and eggs Nest in the city. In spring, the male drums on wood and metal to attract a mate. He then excavates a nest cavity. Female incubates four to six white eggs dawn to dusk and male incubates eggs from dusk to dawn. Although both parents feed insects to nestlings, especially ants, which they dip in sap for extra nutrition, the male brings most of the food to the nest.

Voice High-pitched *kwee-ah, kwee- ah* delivered in sets as one bird calls to another or announces their territory.

Ecological role Omnivores; sap and invertebrates stuck in sap with brushlike tongues. They maintain sap wells every day to keep the sap flowing. This is an important species for many other birds and some mammals. The ruby-throated hummingbird times its migration to the Northeast to that of the sapsucker and uses the sap for nutrition in early spring before flowers bloom.

Yellow-bellied sapsucker holes.

Juvenile yellow-bellied sapsucker at Tanner's Spring, Central Park.

Juvenile yellow-bellied sapsucker feeding on crab apples.

Yellow-bellied sapsuckers working their sap wells.

Young male yellow-bellied sapsucker consuming crab apples, autumn.

Adult male yellow-bellied sapsucker.

Downy Woodpecker: *Picoides pubescens*

Where and when to find The smallest and most widespread of all the woodpeckers, the downy can be seen in trees throughout the five boroughs and at backyard and window feeders, year-round.

What's in a name? *Picoides*: Latin for woodpecker; *pubescens*: hairy; *downy*: referring to the white feathers on the back.

Description Wings and lower back are black with white spots, but its upper back, chest, and belly are white. The outer tail feathers are white with black spots. The head is banded with black and white stripes, and it has white nasal bristles above its bill. The male has a red patch on his head, which the female lacks. Like all woodpeckers, their toes are zygodactyl, and they have stiff retrices. The tiny bill and spotted outer tail feathers distinguish it from the larger hairy woodpecker.

Size 5–7 inches long; wingspan: 10–12 inches.

Behavior The downy woodpecker has stiff tail feathers known as retrices that support it on tree trunks and help it spring forward vertically up the trunk. This is an acrobatic little woodpecker, known for hanging upside down from the end of a branch as it forages for insects, seeds, or fruit. The females stay closer to the trunk.

Nest and eggs Nests in the city. Monogamous during the breeding season, which starts in late winter. After mutual drumming to attract and communicate with a mate, both male and female excavate a nesting cavity in a soft, rotting tree trunk and will often choose a site with lichen or fungus growing on the bark, which helps hide the entrance opening. Both parents incubate four to five shiny white eggs for about 12 days, and both feed the nestlings.

Voice Squeaky, high-pitched *pik* or series of rattling *piks*, ending in a descending whinny.

Ecological role Omnivores; they eat insects plucked from bark crevices and fruit, berries, and acorns. They will come to feeders and consume seeds and suet during winter.

Male downy woodpecker and termite grubs.

Female downy woodpecker. Note the thick, white, nasal
tufts, thought to protect the nasal cavity from pieces of wood
while the woodpecker chisels into tree bark.

Hairy Woodpecker: *Picoides villosus*

Where and when to find Throughout the five boroughs, foraging for insects in the bark of tree trunks in our large and small parks and for seeds and suet at bird feeders during winter.

What's in a name? *Picoides*: Latin for woodpecker; *villosus*: shaggy, named for the long white feathers down its back.

Description Almost identical to the downy, this larger woodpecker's wings and lower back are black with white spots, but its upper back, chest, belly, and outer tail feathers are white. The head is banded with black and white stripes and it has white or tan nasal bristles above its large bill. The male has a red patch on his head. Their toes are zygodactyl.

Size 10 inches long; wingspan: 15 inches.

Behavior Like the downy, the hairy woodpecker has stiff tail feathers known as retrices that support it and help it spring forward vertically up the trunk. Like other woodpeckers, the hairy has a long, barbed tongue that shoots out to catch grubs hiding in the bark; nasal bristles, which protect the eyes from dust and splinters when it is excavating wood; and a thick skull that, like a football helmet, keeps the brain from moving, absorbing the shock of more than 100 blows per minute when it is drumming and excavating tree holes.

Nest and eggs Nests in Central Park and Prospect Park. Monogamous during the breeding season starting in late winter. After mutual drumming to attract and communicate with a mate, both male and female excavate a nesting cavity in a soft, rotting tree trunk. Both parents incubate the three to six glossy white eggs and both feed the nestlings. The young leave the nest after a month.

Voice *Peek* and longer rattling, whinny-like trill. Males emit a flicker-like *wicka-wicka* call when confronting another male.

Ecological role Omnivores; 75 percent of their diet consists of insects, but they also consume fruit, seeds, acorns, and nuts. During early spring, they have been known to drink sap from wells made in the bark by the yellow-bellied sapsucker.

Male hairy woodpecker calling.

Female hairy woodpecker
feeding on small insects
in bark crevices.

Northern Flicker: *Colaptes auratus*

Where and when to find Throughout the city in parks, yards, and woodland edges, spring through autumn.

What's in a name? *Colaptes*: Greek for wood chiseler; *auratus*: Latin for gold.

Description A large woodpecker with beautiful markings: gray head, pinkish-tan face, brown body with black stripes above and black dots below. Red crescent, or chevron, on nape of head, black crescent on top of breast. Males have a black mustache. Wing and tail feathers have golden-yellow shafts. Rumps have white patch that can be seen in flight. They have the long, black, slightly curved beak of an insectivore. Toes are zygodactyl.

Size 13 inches long; wingspan: 21 inches.

Nest and eggs One of the most common nesting birds in New York City. Usually monogamous during breeding season, male and female's courtship dance involves a weaving and bobbing of their heads in opposite directions and fanning of tails as they face each other. Flickers "drum" on trees, roofs, anything metal and loud to attract a mate. Tree cavity nesters; both male and female excavate. The female incubates up to eight shiny, white eggs on a bed of wood chips. Both parents care for the young.

Behavior Parents defend the nest against squirrels and other predators by pecking at them from the nest entrance and attacking them on the tree. The flight of the flicker is like that of all woodpeckers: several strong wingbeats followed by an undulating glide while wings are held close to the body.

Voice *Wicka, wicka, wicka,* also *ki, ki, ki, ki* in rapid repetition; and a piercing *kyeer.*

Ecological role Insectivore; unlike most woodpeckers, the flicker feeds on the ground, probing the soil with its beak for insects, particularly ants, but also beetles, snails, and grubs. They also take insects on the wing, such as butterflies, moths, and flies. Although mainly an insectivore, the flicker feeds on fruit, berries, and seeds if there are not enough insects.

Northern flicker flying over Twin Lakes area, New York Botanical Garden, the Bronx. One of the best areas for birding at the botanical garden. *LM*

Female northern flicker in grass. These members of the woodpecker family often forage for invertebrates on the ground.

Male flicker feeding on crab apples.

Juvenile male and female northern flickers.

Male northern flicker bathing in the Gill, north of Azalea Pond, Central Park.

Male northern flicker holding onto tree cavity with zygo-dactyl toes, two forward and two backward.

THE TYRANNIDAE family, with more than 400 species, is the largest family of birds on Earth. All of the flycatchers in this family live in the Western Hemisphere of North and South America, with the largest diversity of species living in South America.

Except for the vermillion flycatcher and the colorful ornate flycatcher, most species in this family are brown, gray, and white. They have large heads and flat bills with conspicuous bristles at the base, short legs and feet, and crown feathers that can be raised to attract mates or ward off intruders. Many are aggressive; hence, the family name Tyrannidae: tyrant.

In New York City, members of this family include the eastern wood-pewee, the eastern phoebe, and the eastern kingbird. All three species hawk for insects from a tree perch. The kingbird, *Tyrannus tyrannus*, is particularly aggressive.

EASTERN PHOEBE

EASTERN
WOOD-PEWEE

EASTERN
PHOEBE

EASTERN
KINGBIRD

Eggs shown at life size.

Eastern Wood-Pewee: *Contopus virens*

Where and when to find This tiny bird migrates to the city from South America in early May and can be found in city parks from May through June and August through September.

What's in a name? *Contopus*: Latin for short footed, referring to the small toes; *virens*: Latin for green, referring to the olive tint of its back and head; *wood:* found in deciduous woods; *pewee*: its call.

Description Upper parts are grayish olive, with a darker crown and nape. The throat and belly are light gray or white, with olive to gray bands on the chest and down the sides. Wing feathers are edged in tan, with two whitish wing bars. Tail dark brown with a notch in the middle. Bill is dark above and yellowish orange below, with bristles above, along the sides, and below the bill.

Size 6 inches long; wingspan: 10 inches.

Behavior Perches halfway up a tree on the tip of a bare branch, from which it flies out to "hawk" for flying insects. When feeding their young, they make almost 70 foraging flights an hour. When they are not breeding, they make 36 feeding flights per hour.

Nest and eggs Nests in Van Cortlandt Park. Female builds a small nest and binds it to the branch with spider's silk and plant fibers 15–60 inches above the ground. She uses grass and weed stems, shredded bark, thread for the outer cup and lines it with fine grasses, plant down, and animal hair. She decorates the outside, which camouflages the nest, with spider webs and lichen so that the nest looks like a knot on the branch. She lays three white or cream-colored eggs speckled with flecks of brown, purple, or both, which wrap, garland-like, around the larger end of the egg.

Voice *Pee-ah-WEE, pee-ah-WEE.*

Ecological role Insectivore; it feeds on all small flying insects.

Eastern wood-pewee, crown feathers slightly raised, on yew twig.

Eastern wood-pewee; note the wide base of its flycatcher bill.

Eastern Phoebe: *Sayornis phoebe*

Where and when to find In early spring, when it migrates to city parks and backyards, into autumn when it migrates south.

What's in a name? *Sayornis*: Latin for Say's bird. Thomas Say, an American naturalist, 1787–1834; *ornis:* Latin for bird. *Phoebe:* the bird's call.

Description A medium flycatcher with a brown head, face, back, and tail. Its head is flat on top and appears large. The beak is short and thin for capturing insects. Flycatchers as a group have this type of bill. With dark washes along its sides, the breast and belly are white, which makes the brown face and head stand out, a good diagnostic tool for identifying the bird.

Size 7 inches long; wingspan: 10.2–11 inches.

Behavior A true harbinger of spring. If you see a brown and white bird in late March flicking its tail, you'll know it's a phoebe, returning to the city to breed. Nesting on buildings and human-made structures has allowed them to adapt to such habitat changes as the loss of cliff ledges. They will migrate back to their old nests year after year and fix them up, adding new mud and moss. If they abandon an old nest, they usually build a new nest near it under the same overhanging structure.

Nest and eggs Nests in city. Female builds the nest under a porch, a deck, or a building eave for protection. She builds the outer nest with mud and dried grass, covers it in moss, and lines it with fine plant fibers, rootlets, and hair. I have seen a video of a phoebe gathering hair from a golden retriever sleeping on a porch.

Voice *FEE-bee, FEE-bee* uttered over and over. The voice of the fledglings is a raspy, thin *FEE-bee.*

Ecological role Insectivores; feed on bees, wasps, beetles, butterflies, grasshoppers, and flies. Their bills are wide and flat with hairy bristles at the base, which helps them channel insects into their mouths. They "hawk" for insects, catching them midflight but will also glean them from plants.

Eastern phoebe, typical pose on fencepost. Note rictal bristles, which are thought to help flycatchers hold onto prey.

Eastern phoebe feeding on sulphur butterfly.

Eastern Kingbird: *Tyrannus tyrannus*

Where and when to find May through September, in city parks and backyards in all five boroughs.

What's in a name? *Tyrannus*: Latin for tyrant, referring to its pugnacious nature; kingbird: ruler of his territory.

Description Black head with a concealed crown of red, yellow, and/or orange, the kingbird has a dark gray back and wings. Its long, black tail has prominent white tips. The throat, chest, and abdomen are white.

Size 7.5–9.1 inches long; wingspan: 13–15 inches.

Behavior Can be quite aggressive with large birds, including hawks and crows that fly too close to its nest, and will raise its crest and flare its crown as it flies after the intruder. Has been known to fly into blue jays and knock them out of trees. During spring and summer, it feeds on insects. It can hover over the ground and then swoop down to capture insects in midair or on the ground.

Nest and eggs Mate for life. Both male and female build their nest close to the end of a horizontal tree branch using twigs, weed stems, dried grass, hair, feathers, and string and line it with rootlets, fine dried grass, and hair. Female incubates three to five slightly glossy, creamy-white or pinkish eggs spotted with reddish brown, purplish brown, blackish purple, pale brown, lilac, and gray markings, typically around the larger end of egg. Parents care for the young for more than a month.

Voice Male sings *t't'tzeer, t't'tzeer, t'tzeetzeetzee* continuously early morning and late afternoon through early evening.

Ecological role Omnivores; feed on insects during warm seasons: beetles, ants, wasps, bees, crickets, flies, and grasshoppers. Often seen perched out in the open, on treetops, wires, or fences; hawks for insects, catching them on the wing. In late summer and early fall, will feed on fruit: mulberries, serviceberries, elderberries, cherries, blackberries, and nightshade berries.

Eastern kingbird feeding on cicada killer wasp.

Eastern kingbird perching in sun. Notice the white tail tips, a good field mark for identifying kingbirds.

THERE ARE 50 SPECIES of vireo, all native to the new world: North and South America. They are small, with short, thick bills, slightly hooked at the tip, which they use to hawk insects in midair and to glean them from leaves. Birds of the woods, their plumage is typically dull brown, yellow, and green above and pale below. Eye stripes or white spectacles are common around their eyes. They feed on invertebrates and sometimes fruit.

In North America, there are 13 species of vireo: Bell's, warbling, white-eyed, black-capped, gray, yellow-throated, plumbeous, Cassin's, blue-headed, Hutton's, Philadelphia, red-eyed, and black-whiskered. They typically forage for insects high in the treetops. One can usually hear them, but it is not easy to see them.

In New York City, the most common vireo is the red-eyed. It is quite a songster, singing continuously as it forages for insects high in the trees of city parks and backyards. Other vireos seen in the city are the warbling, white-eyed, blue-headed, and Philadelphia. These species can be seen in spring and fall in parks along the coast, such as Mount Loretto in Staten Island, Breezy Point in the Rockaways, Jamaica Bay Wildlife Refuge, Floyd Bennett Field, Fort Tilden, and Jacob Riis Park, and in inland parks: Forest Park in Queens, Prospect Park in Brooklyn, Central Park, Fort Tryon Park, Inwood Hill Park in Manhattan, and Van Cortlandt Park and Pelham Bay Park in the Bronx.

RED-EYED VIREO

Egg shown at life size.

Red-eyed Vireo: *Vireo olivaceus*

Where and when to find You will hear its beautiful song first and might have to use binoculars to see this bird high up in trees, often hidden by leaves. It migrates to the city in early May and stays until mid-October in city parks and in areas with large trees.

What's in a name? *Vireo*: Latin for green; *olivaceus*: Latin for olive green, referring to color of bird's upper parts.

Description A tiny bird of the tree canopy; it has a gray or blue-gray crown, olive upperparts, and gray to white chest and abdomen. It has a white "eyebrow," with a dark stripe across its red eyes.

Size 4.7–5.1 inches; wingspan: 9.1–9.8 inches.

Behavior One of the few birds that sings in the middle of the day. It sings continuously from dawn to dusk, high in the trees. The males are among the most tenacious and prolific singers in the bird world. In spring, they have been recorded singing more than 20,000 songs in 10 hours. They move along branches high in the tree as they search for prey. They kill larger insect prey by beating it against the branch and hold it with their foot while eating.

Nest and eggs Nests in the city. Female builds a hanging nest and binds the rim to a forked branch with spider's silk or caterpillar web. She uses vine bark, fine grasses, rootlets, and birch bark to build the cup and will sometimes decorate it with lichen. She incubates three to five smooth white eggs, speckled with small reddish-brown, brown, or black dots. Both parents care for the young.

Voice Song is a series of repeated phrases and pauses, *look up . . . see me?. . . over here . . . this way . . . do you hear me?. . . higher still . . .* All these phrases are sung with a rising note at the end, and with a pause, as if waiting for an answer.

Ecological role Omnivores; consume large quantities of insects, especially the destructive caterpillars of gypsy moths and fall webworms. They eat fruits in winter.

Red-eyed vireo, looking eager and alert.

Red-eyed vireo perched on pokeweed.

JAYS AND CROWS

THE CORVIDAE family includes jays, crows, ravens, magpies, nutcrackers, choughs, and treepies. Worldwide there are roughly 120 species in every habitat but the polar regions and southern portion of South America. They are the most intelligent of all birds, and recent research compares their intelligence with that of apes and just below humans. They have a language, can count, use tools, solve puzzles, form cooperative family groups, and display amazing memory skills.

Corvids are medium to large songbirds, usually with black or blue plumage, sometimes they are pied with black and white feathers. South American members of this family are more colorful. The beautiful Sri Lankan magpie is blue and red. Corvids have stout bodies, strong legs, long, thick bills with tufts of feathers around their nostrils, and rictal bristles around their mouths. Some species, such as the blue jay, American crow, and raven, mate for life. Their calls are loud and harsh. Most species are thriving but some, such as the Sri Lankan magpie, are declining because of habitat destruction.

Corvids are omnivores and are generally nonmigratory. They will stay in their home territory year-round and for generations, migrating only if there is a shortage of food. American crows, fish crows, magpies, ravens, nutcrackers, and many species of jays live in North America. In New York City, there are occasionally fish crows and ravens and abundant blue jays and American crows.

BLUE JAY

BLUE JAY · AMERICAN CROW

Eggs shown at life size.

Blue Jay: *Cyanocitta cristata*

Where and when to find Can be heard and easily seen in city parks and street trees in all five boroughs, year-round.

What's in a name? *Cyanocitta:* chattering blue bird; *cristata:* crest.

Description A beautiful bird, with a brilliant blue crest, nape, back and tail, often with shades of purple. The tail and wings have black barring and white spots. Their face, chest, and belly are white. They have a black "necklace" extending up the sides of their face to their crest. Their feathers are actually brown. The blue color is created by sunlight refracted from the feathers, scattering blue wavelengths.

Size 10–12 inches long; wingspan: 13–17 inches.

Behavior Like their relative the crow, blue jays are intelligent and curious birds. They raise their crests when alarmed and lower them when they are with family members or feeding their young. They defend the nest against predators by mobbing hawks, owls, raccoons, squirrels, and humans. Although they are thought to be aggressive toward other birds, only 1 percent feed on birds and their eggs.

Nest and eggs Blue jays bond for life and are devoted mates. Both build the nest; the male gathers most of the twigs, grasses, and mud, and the female arranges them. The female incubates and broods four to six pale buff-bluish eggs with brown spots, while the male feeds insects to her and the nestlings when they hatch.

Voice Blue jays have an extensive number of vocalizations due to a Y-shaped syrinx, which produces two sounds at once. The most commonly heard is the alarm call *jay jay jay*. Another mild alarm call is a soft, bell-like *tull lull*. They mimic calls of hawks when agitated. One theory is that this signals other birds that a hawk or other aggressor is near. Another theory is that the blue jay's hawk cries will scatter other birds from a feeder.

Ecological role Omnivores; feed on insects, seeds, nuts, and berries. Blue jays are valued for helping the growth of oak forests due to the "forgotten" acorns they bury and for consuming harmful insects such as Japanese beetles and tent and gypsy moth caterpillars.

A bonded pair of blue jays, who mate for life.

Blue jay collecting muddy nesting material at Tanner's Spring, Central Park.

Blue jay feeding nestlings.

Older blue jay nestling calls for food, mouth agape.

Blue jay swallowing peanut just removed from its shell in snow.

The blue jay's blue color is produced by the structure of its feathers, which refract blue light. If feathers are backlit, they appear brown.

American Crow: *Corvus brachyrhynchos*

Where and when to find Throughout the five boroughs, year-round: we usually hear crows before we see them flying overhead.

What's in a name? *Corvus*: Latin for raven. *Brachyrhynchos*: ancient Greek for short billed, meaning that the beak of the crow is shorter than the beak of the raven.

Description Large bird with glossy, black feathers that show a violet iridescence when birds are mature (older than 15 months). Their eyes, bills, legs, and feet are also black. The tail is rounded. Males are slightly larger than females.

Size 17–21 inches long; wingspan: 33.5–39.4 inches.

Behavior Among the most intelligent of birds; they can shape and use twigs and leaves as tools. They mob predators, such as hawks, that come near their nest by attacking with their bodies and loud vocalizations.

Nest and eggs Crows mate for life. Both parents build the large, sturdy nest, often with the help of their offspring from previous years, who bring sticks and leave them for the female to arrange and line with weeds, bark, fur, and mud. Female incubates four to six pale blue to green eggs with blotches of brown and gray. The male and family helpers bring food to the female and the hatchlings.

Voice A series of *caw*s; duration of each *caw* and the amount of time between *caw*s communicate information.

Ecological role Omnivores; consume invertebrates and vertebrates, including frogs, tadpoles, toads, salamanders, small birds, bird eggs, and mammals, such as young squirrels, mice, carrion, and human food. Their cousin, the similar-looking fish crow—hit hard by West Nile virus—also nests and lives in New York City.

American crows attacking squirrel in tree den. As crows do not typically feed on squirrels, they were probably defending their nest.

Crow discovering stashed acorn in tree hole.

Crow with its acorn.

SWALLOWS AND MARTINS belong to the Hirundinidae family. *Hirundo* is Latin for swallow. There are approximately 80 species of birds in this family worldwide, with most species living in Africa. Though they can live in a wide diversity of habitats, they are usually found near wetlands, hunting for insects flying over water. Swallows and martins spend more time in flight than any other songbird.

With small, slender bodies and long, pointed wings, they are incredibly acrobatic, flying so low to feed, they appear to touch the water or ground as they capture prey, often in midair. They drink from the surface of the water as they are flying. Their upper body feathers are dark and iridescent. Their northern migration in spring is timed to coincide with the appearance of insects, and swallows are hailed as the harbingers of spring.

The following species live in North America: the bank swallow, barn swallow, cave swallow, northern rough-winged swallow, tree swallow, violet-green swallow, and purple martin. In New York City, the barn swallow and tree swallow are commonly seen in spring and summer, flying low over the meadows, ponds, and lakes of our city parks, scooping insects out of the air. A purple martin colony returns to their nests in Staten Island along Lemon Creek. The nest boxes were established in 1952 by Howard Cleaves, who worked for the Staten Island Museum. The museum and local residents help maintain this purple martin sanctuary.

BARN SWALLOW

TREE
SWALLOW

BARN
SWALLOW

Eggs shown at life size.

Tree Swallow: *Tachycineta bicolor*

Where and when to find Throughout the five boroughs, wherever there are abundant flying insects: over ponds, lakes, fields, and parking lots near water, from April through October.

What's in a name? *Tachycineta*: Latin for swift mover, referring to its fast flight; *bicolor*: Latin for two colors, referring to the dark upper parts and white lower parts; *tree*: nest in trees; *swallow*: Anglo Saxon word for this bird is *swalewe*.

Description Tiny birds, with pointed wings and slightly notched, sometimes square, "swallow" tails. Males are iridescent blue green above and white below. Females are duller brown above, and juveniles are completely brown above.

Size 4.7–5.9 inches long; wingspan: 11.8–13.8 inches.

Behavior Acrobatic fliers, twisting and turning in midair at 20 miles per hour, they chase and capture flying insects. They are easily recognizable, with white throat, chest, and abdomen and iridescent blue and green upper parts glistening in the sun.

Nest and eggs Abundant nesters in Jamaica Bay Wildlife Refuge because of the many nest boxes put up in the refuge by guardian Don Riepe. They do nest in natural tree cavities and woodpecker holes but will happily move into nest boxes. A pair of tree swallows made a nest inside the exhaust pipe of a boat at the 79th Street Boat Basin on the Hudson River each spring, until the owner built a nest box for them, which they immediately used. The female builds a nest of dried grasses and pine needles and will line the cup with feathers. She incubates four to six tiny, pure white eggs, less than an inch long. Both parents feed the nestlings.

Voice male sings his dawn song: *teet, trrit, teet, teet, trrit, teet, trrit.* Both male and female sing a short series of high notes followed by a liquid gurgle.

Ecological role Insectivores; a natural pest controller, they consume great numbers of flies, grasshoppers, beetles, moths, bees, and wasps. In autumn, before migration, if insects are scarce, they feed on seeds and berries.

Female tree swallow, Central Park.

Tree swallows at their nesting box, West Pond, north side of loop,
Jamaica Bay Wildlife Refuge. *LM*

Barn Swallow: *Hirundo rustica*

Where and when to find Throughout the five boroughs, wherever there are abundant flying insects: over ponds, lakes, fields, and parking lots near water, from April through September.

What's in a name? *Hirundo*: Latin for swallow; *rustica*: rural *swallow*: Anglo Saxon word for this bird is *swalewe*.

Description Blue back wings, tawny-rufous underparts, and a long, deeply forked blue tail. Blue crown and face against a cinnamon forehead and throat and a short, wide beak. Males are more colorful than females.

Size 5.9–7.5 inches long; wingspan: 11.4–12.6 inches.

Behavior An extremely graceful aerialist; it swoops and darts over fields, yards, and open water in search of flying insects. Often cruises low, flying inches aboveground or over water. It feeds and drinks by skimming over ponds and lakes and scooping up insects or water in its open mouth.

Nest and eggs Both male and female build the nest. They collect mud and mix it with grass stems to make pellets, which they use to cover the outside of the nest. Both parents take turns incubating four to five tiny, glossy-white eggs speckled in lilac, gray, and reddish brown. Nestlings are fed by male and female, but sometimes they get help from older juveniles of previous clutches.

Voice Sings from April through August, a long, bubbly and chirping song.

Ecological role Insectivores; the barn swallow is beneficial in that it feeds on harmful insects, including houseflies and mosquitoes.

Barn swallow on wire at Pier I, Hudson River, and 70th Street.

Barn swallow mud nest lined with dry grass and feathers. *LD*

CHICKADEES, titmice, and their allies, nuthatches and creepers, belong to several bird families.

The Paridae family includes titmice and chickadees. *Parus* is Latin for titmouse, *tittr* is old Icelandic for small, and *mase* is old English for bird. Small and stocky, they feed on seeds and insects. They are acrobatic feeders, often hanging upside down on seedpods. They work tirelessly to extract seeds from hulls, often holding the seed down with both feet and hammering it open. Sixty species of Paridae live in North America, Europe, Asia, and Africa. In North America, there are three species of chickadees (black-capped, Carolina, and boreal) and three species of titmice (tufted, juniper, and black-crested).

Nuthatches belong to the Sittidae family. *Sitte* is Greek for nuthatch. Up to 30 species live in Asia, North America, and Europe. They are small birds with large heads, short tails, and feet with long talons for grasping tree trunks as they glean insects using long, thin bills. Acrobatic tree climbers, they often forage moving face down on trunks. There are four species of nuthatches in North America—red-breasted, white-breasted, pygmy, and brown-headed.

Creepers belong to the Certhiidae family, which include 11 species in North America, Europe, Asia, and sub-Saharan Africa. *Certhius* means creeper in Latin. In North America, there is one species, the brown creeper. Creepers are small, streaked brown, with thin sharp bills for extracting invertebrates from bark crevices. They creep around tree trunks for food.

In New York City, black-capped chickadees, tufted titmice, white-breasted nuthatches, and brown creepers are common. Less common, but sometimes seen, are red-breasted nuthatches.

TUFTED TITMOUSE

BLACK-CAPPED
CHICKADEE

TUFTED
TITMOUSE

WHITE-
BREASTED
NUTHATCH

BROWN
CREEPER

Eggs shown at life size.

Black-capped Chickadee: *Poecile atricapillus*

Where and when to find Year-round in the wooded areas of city parks and bird feeders in all five boroughs.

What's in a name? *Poecile*: variegated black-and-white color pattern; *atricapillus*: black crown.

Description A tiny gray bird with a black cap, black bib, long gray tail, white chest and abdomen, and white cheeks. The bright white cheeks separating the black cap and bib are a good field mark from a distance.

Size 4.75–5.75 inches long; wingspan: 6.3–8.3 inches.

Behavior Chickadees are extremely tame. They fly to feeders and take one seed at a time. Holding the seed between their feet, they hammer it with their beaks until they crack it open and extract tiny bits of seed. They are well known to hide seeds and can remember hiding places.

Nest and eggs Nest in the city. Monogamous, their pair bond can last until one dies. Pairs establish and defend a territory, remaining on or near their territory for life. Black-capped chickadees nest in cavities they excavate themselves and construct a nest cup within the cavity made of moss and often lined with fur, plant down, feathers, and cocoons. The female incubates up to 10 white eggs speckled with reddish-brown spots. Both parents feed the young.

Voice With 16 different vocalizations, the most easily recognizable is *chick-a-dee dee dee*. The more *dees*, the more urgent the call. Other species of birds foraging near a flock of chickadees have learned to heed this threat call. *Fee-bee* is the male's song during the breeding season.

Ecological role Omnivores; feed on invertebrates such as spiders, insects, millipedes, and mollusks, including slugs and snails. They also feed on seeds, berries, and nuts.

Black-capped chickadee perched on twig.

Black-capped chickadee foraging for seeds, hanging upside down on dead leaf.

Tufted Titmouse: *Baeolophus bicolor*

Where and when to find Can be seen through the four seasons in the wooded areas of the city's parks in all five boroughs.

What's in a name? *Baeolophus*: small crest; *bicolor*: two colors; *tufted*: crested; *titmouse*: Old English for tiny bird.

Description Small, crested bird. Both males and females have a characteristic black forehead and gray crest. The upperparts are gray, with darker gray flight feathers. The face and underparts are white with amber flanks. The dark eye and eye ring are conspicuous on the white face.

Size 5.5–6.3 inches long; wingspan: 7.9–10.2 inches.

Behavior In winter, small flocks of titmice appear in mixed flocks with black-capped chickadees and white-throated sparrows foraging for seeds on the ground. They hang upside down exploring pinecones and seed-pods. Holding a seed between their feet, they hammer it open with their beak. Small winter flocks of titmice are composed of parents and their young. They are tame and intelligent birds.

Nest and eggs Breed during spring and summer. Pairs mate for life. Titmice build their nests in tree cavities made by woodpeckers, fungus, and other decomposing organisms. The female usually lays five to seven white eggs with brown spots. Eggs take about two weeks to hatch. Both parents feed their young.

Voice *Peter peter peter.*

Ecological role Omnivores; titmice feed on a wide variety of foods, including spiders and their egg cases, caterpillars, wasps, bees, scale insects, ants, beetles, treehoppers, other insects, snails, acorns, beechnuts, blackberry, elderberry, blueberry, grape, serviceberry, ragweed, sunflower seeds, pine seeds, Virginia creeper berries, hackberries, and mulberries. They eat more animal protein during the warm seasons and nuts, berries, and seeds in winter.

Tufted titmouse perched on twig on ground in leaf litter.

Tufted titmouse opening a bagworm cocoon to get to the
caterpillar larvae within.

White-breasted Nuthatch: *Sitta carolinensis*

Where and when to find Rare in summer; otherwise, found in the wooded areas of the city's parks in all five boroughs.

What's in a name? *Sitta*: Greek for a bird that pecks at trees; *carolinensis*: from Carolina where bird was identified and named; *nuthatch*: using the bill to hack open nuts.

Description Both the male and female white-breasted nuthatch have a blue-gray back with a white throat, breast, and belly. Their dark eyes stand out against their unmarked white face. Males have a black cap and nape; females and juveniles have a gray cap and nape. Their black tail feathers are patterned with white bands. The central tail feathers are pale bluish gray. Undertail coverts are white marked with reddish orange. The long, upturned bill is dark above, light below, and adapted for probing bark crevices and for hacking open seeds and nuts.

Size 5–6 inches long; wingspan: 7.9–10.6 inches.

Behavior Collects insects from the bark of trees, usually creeping head first down the trunk. They work their way down a tree trunk searching for grubs and insects in the bark crevices that other up-climbing tree foragers, such as the woodpecker, would miss. In winter, it may feed in flocks with chickadees and titmice. At feeders, the nuthatch often picks up one seed at a time and flies away with it to a tree where it hacks it open with its beak.

Nest and eggs Nest in Staten Island. They mate for life and are cavity nesters, using old woodpecker holes and natural tree cavities. The female gathers nesting material such as bark and lines the cup with feathers. She lays about five to eight creamy white eggs speckled with reddish-brown or purplish-red spots. The female incubates the eggs and the male brings her food. Both parents feed their young.

Voice It often calls *yank, yank* while feeding.

Ecological role Omnivores; feed on insects and spiders during the summer and nuts, berries, and seeds in the winter, which is also when they store food in the bark crevice of tree trunks.

White-breasted nuthatch in typical pose headed down the tree trunk.

White-breasted nuthatch showing slightly upturned bill adapted
for probing bark crevices and hacking open seeds.

Brown Creeper: *Certhia americana*

Where and when to find Sparsely wooded areas of city parks and yards with large trees. Mid-September through April in all five boroughs; April and October along the coast.

What's in a name? *Certhia*: Latin for creeper; *americana*: of America; creeper: its habit of creeping around tree trunks searching for food.

Description Head and back feathers are brown dappled in white; underparts are all white, wide tan stripe over eyes, and fluffy white throat. Bill is long, extremely slender, and decurved. Wings have pale-yellow stripes and the stiff, brown tail, which is used as a prop, has a rufous rump.

Size 4.7–5.5 inches long; wingspan: 6.7–7.9 inches.

Behavior A tree clinger as well as a creeper, this tiny bird has the habit of creeping around the largest tree trunks and branches. The slender downcurved bill is adapted for probing bark crevices and tearing away loose bark in search of insect prey. They begin at the base of a tree, spiral around the trunk foraging beneath the bark until they get to the top of the tree. Then they fly to the base of a nearby tree and resume their foraging.

Nest and eggs In 1879, naturalists discovered that they build their hammock-shaped nests behind peeling bark, often in dead trees. They never fly directly to the nest, but rather they land at the base of the tree and creep up the trunk. They will also build their nests behind a screen of ivy or other leafy vines. Nest is made of twigs, woody fibers, grass, moss, and roots. It is lined with feathers and finely shredded bark. Female typically incubates six white eggs, speckled with reddish-brown spots. Both parents care for the hatchlings.

Voice Call is very high, short, often a piercing *see* or *swee*. Song: *pee pee willow wee* or *see tidle swee.*

Ecological role Insectivores; glean invertebrates, mostly insects and spiders, from crevices in the bark.

Brown creeper feeding on white moth.

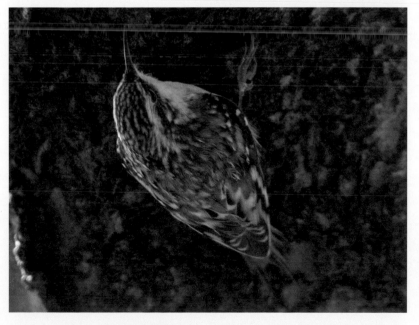

Brown creeper clings to tree in search of insects. Its long decurved bill is adapted for tearing away loose bark in search of prey.

THERE ARE 80 SPECIES of wrens in the Troglodytidae family. This family name is from the Greek word for one that creeps into holes, referring to the wren's practice of using cavities for nests and creeping around crevices searching for invertebrates. Except for the Eurasian wren, all wren species live in the Western Hemisphere, primarily in Central and South America.

There are nine species of wrens in North America: sedge wren, marsh wren, house wren, winter wren, Carolina wren, Bewick's wren, rock wren, canyon wren, and cactus wren.

Most wrens are tiny, speckled or streaked gray and brown birds with slender, sharp decurved bills for plucking insects and tails that stand straight up at times. They forage for invertebrates in the leaf litter. They have beautiful songs and can sing for hours.

In New York City, the house wren and Carolina wren are commonly seen—the house wren in the spring and summer but the Carolina wren increasingly year-round. Marsh wrens and winter wrens are seen in some city parks.

CAROLINA WREN

HOUSE
WREN

CAROLINA
WREN

Eggs shown at life size.

House Wren: *Troglodytes aedon*

Where and when to find April through October, abundant in parks and back-yards of all five boroughs.

What's in a name? *Troglodytes*: Latin for one who creeps into holes, referring to cavities it nests in; *aedon*: Latin for nightingale; house: nests near homes; wren: Anglo Saxon: *wraenna*: for this type of bird.

Description Small, with finely striped brown head, nape, back, and tail, with a pale stripe above the eyes. Throat and chest are light gray and flanks have pale brown stripes. Bill is long and slightly curved at the tip. Tail held straight up.

Size 4.3–5.1 inches long; wingspan: 6 inches.

Behavior Hop on the ground while foraging for insects in the leaf litter and fly about 3 feet off the ground in open areas as they hunt for invertebrates. Can be aggressive and territorial with other cavity nesting birds. The male will crouch, drop his wings, erect his back feathers, and lower his fanned out tail.

Nest and eggs An eccentric nester, known for choosing unusual cavities such as the top of a hat or a tin can as well as traditional woodpecker holes and nest boxes. The male fills the site with large twigs, leaves, plant fibers, and grass. The female lines the nest with fine grasses, feathers, hair, wool, and spider eggs. When spiderlings hatch, they feed on destructive nest parasites. Wrens leave a small opening that they can use, which is too small for a larger bird. The female incubates six to eight glossy white eggs, which look pink because they are finely speckled with purplish-red or purplish-brown spots. Both parents feed the young.

Voice Male and female sing a rich, bubbling song that starts slowly, rises in volume and pitch, and then drops and ends with a rapid series of cascading notes.

Ecological role Insectivores; consume beetles, caterpillars, earwigs, daddy longlegs, flies, leafhoppers, and springtails. Sometimes they feed on snail shells for calcium and on grit, which helps digest their food.

House wren and its nest atop Central Park lamppost.

House wren singing from its tree perch. Males advertise for females from trees; otherwise, they sing on the ground.

Carolina Wren: *Thryothorus ludovicianus*

Where and when to find Year-round in woody areas of city parks and backyards of all five boroughs.

What's in a name? *Thryothorus:* Latin for rushing; referring to its habit of running through underbrush as it forages for food; *ludovicianus:* Latin for King Louis XIV of France; named for the Louisiana Territory where it was first collected; *Carolina:* general term for southeastern United States, where it was commonly found; wren: Anglo Saxon; *wraenna:* meaning wren.

Description Warm, rich cinnamon head and back with small dark stripes on wings and tail. The throat is whitish, the chest and abdomen a paler cinnamon, a bright-white eyebrow, and a long, curved bill.

Size 4.7–5.5 inches; wingspan: 11.4 inches.

Behavior Pairs mate for life and remain in the same area year-round. It holds its tail up while foraging. Male generally keeps tail down when singing. They forage in leaf litter, tangled vines, and beneath fallen trees. Global warming has allowed this species to move north since the turn of the twentieth century. It does well in mild winters but not in extreme cold with snow and ice. However, within a few years, its northern numbers rebound.

Nest and eggs Nests in city in a variety of natural cavities. Male builds a domed nest made of grasses, stems, bark, leaves, spider webs, rootlets, and some moss. The female lines it with fine grasses, rootlets, hair, and feathers. Female incubates four to eight white eggs speckled with brownish red and purplish brown.

Voice *Tea-kettle, tea-kettle, tea-kettle! che-wortel, che-wortel, che-wortel,* and *choo-wee, choo-wee, choo-wee.* Male sings to defend territory and can sing thousands of times a day.

Ecological role Omnivores; feed on spiders, caterpillars, moths, crickets, beetles, grasshoppers, cockroaches, and a small amount of plant matter and seeds from bayberry, sweet gum, poison ivy, sumac, and smartweed.

Male Carolina wren singing. Only males sing and have a larger, more developed song region in their brains than females.

Carolina wrens often hold their tails straight up.

FORMERLY CONSIDERED PART of the old world warbler family Sylviidae, kinglets now have been placed in the family Reguliidae, from the Latin *regulus*: little king, referring to their colorful crown patches. There are six species: the goldcrest of Europe and Asia, the colorful common firecrest of southern Europe and North Africa, the Madeira firecrest of Portugal, the flamecrest of Taiwan, and the two North American species— the ruby-crowned and golden-crowned kinglets.

These six species are the smallest of the songbirds, between 3 and 4 inches long, with tiny, tweezer-shaped bills for grabbing insects. Males and females are alike, but males have a brighter crown patch that can be raised during courtship and when defending their territory.

The kinglets are among the first migrants to return to New York City in spring, sometimes in mid-March while there is still snow on the ground. They typically pass through in small flocks of 8 or 10, foraging for insects.

RUBY-CROWNED & GOLDEN-CROWNED KINGLETS

GOLDEN-
CROWNED
KINGLETS

RUBY-
CROWNED
KINGLETS

Eggs shown at life size.

Golden-crowned Kinglet: *Regulus satrapa*

Where and when to find In all five boroughs during spring migration, late March through April, and again in fall migration, October through November.

What's in a name? *Regulus*: Latin for king; *satrapa*: wealthy ruler.

Description Tiny, gray, with olive-yellowish-green shoulders and wings with white and black wing bars, pale gray below. Males have yellow-orange crown within black border. The crown turns a deep orange during breeding season. Female has pale-yellow crown. They have a white bar above and below their eyes and a black bar across the eyes.

Size 3.1–4.3 inches long; wingspan: 5.5–7.1 inches.

Behavior Flight is quick and fitful. They often flick their wings. They can hang upside down from a twig and hover while feeding. They move in small groups, gleaning tiny insects from leaves and twigs. They raise their golden crowns while displaying to attract a mate or when agitated.

Nest and eggs They breed in the conifer forests of the northern United States and Canada but not in New York City. Nests are built by male and female on evergreen trees, consisting of materials that can expand as the hatchlings grow. They collect needles, spider silk, grasses, cotton, deer hair, feathers, lichen, and mosses. Tree bark gives it structure. They suspend their nest from the ends of branches, beneath leaves. Female lays eight or nine eggs. Male feeds her and the hatchlings.

Voice Call: high pitched one to five *tsees*; song is a complex series of up to 14 ascending *tsee* notes ending in a trill.

Ecological role Carnivores; glean insects, spiders, mites, and invertebrate eggs from trees and shrubs, flower buds, and beneath tree bark.

Migrating golden-crowned kinglet after snowstorm, Halloween weekend, 2011.

Golden-crowned kinglet perched in flowering smartweed patch.

Ruby-crowned Kinglet: *Regulus calendula*

Where and when to find In parks of all five boroughs, feeding on tiny invertebrates on shrubs and trees during spring migration, late March through April, and again in fall migration, September through December.

What's in a name? *Regulus*: Latin for king; *calendula*: Latin for glowing, referring to the red crown patch.

Size 3.5–4.3 inches long; wingspan: 6.3–7.1 inches.

Behavior Flits from branch to branch, gleaning invertebrates. Constantly flicks its wings, a characteristic that helps distinguish it from the golden-crowned kinglet.

Nest and eggs Breeding occurs in the far north in conifer forests. The spherical nest may be resting on the branch of a tree, or suspended from it, sometimes 100 feet high and typically concealed by foliage. The female builds a nest using grass, moss, spider and cocoon silk, and feathers. She lines it with animal fur and soft plant material and lays 5–11 white eggs with reddish-brown spots around the larger end. This is the largest number of eggs laid by any North American songbird. The spider webs and cocoon silk make the nest stretch as the hatchlings grow. Both parents feed the young.

Voice Extremely loud for its size, the notes sound like *zee-zee-zee* followed by several low trills, and in males, the song ends with *tee-da-leet, tee-da-leet.* Song described as one of the richest of all the birds.

Ecological role Omnivores; feed on invertebrates, such as spiders, aphids, wasps, beetles, and ants, and occasionally feeding on fruit such as poison ivy berries and dogwood berries.

Male ruby-crowned kinglet; red crown, normally hidden, is exposed when it's agitated.

Ruby-crowned kinglet feeding on a hornet.

THRUSHES belong to the Turdiidae family. There are approximately 125 species worldwide, with most living in Europe and Asia. Thrushes are insectivores that may also feed on berries and fruit. They have round bodies and are small to medium size. They often have colorful plumage, with blue, orange, and white feathers. Except for the eastern, western, and mountain bluebirds, the North American thrushes have brown, gray, or black feathers above and spotted breasts. The American robin is brown above and orange below, but juvenile robins have the typical throat spots of the thrush.

Thrushes typically live and nest in trees, usually feeding on invertebrates on the ground. They are insectivorous, but most species also eat worms, land snails, and fruit. Many species permanently reside in tropical climates, while others migrate in spring to their northern breeding grounds where they build cup-shaped, mud-lined nests in trees. Thrushes are loved for their songs, some of the most beautiful in the bird world, such as the songs of the North American hermit thrush and the Eurasian nightingale.

There are 14 species of thrushes in North America. The eastern bluebird, western bluebird, mountain bluebird, hermit thrush, wood thrush, gray-cheeked thrush, Bicknell's thrush, Swainson's thrush, American robin, veery, Townsend's thrush, fieldfare, clay-colored robin, and the varied thrush.

Commonly seen and nesting in New York City's parks and backyards are the hermit thrush, wood thrush, veery, and the American robin. In many parks, you can find American robins in the city year-round, as they feed on berries throughout the winter.

AMERICAN ROBIN

VEERY   HERMIT THRUSH   WOOD THRUSH   AMERICAN ROBIN

Eggs shown at life size.

Veery: *Catharus fuscescens*

Where and when to find Moist, wooded areas of city parks during May when they stop over on their way north to breed, and in September, when they are migrating south.

What's in a name? *Catharus*: Latin for pure, referring to the pure notes of its song; *fuscescens*: Latin for somewhat dark colored; *veery*: the sound of its song.

Description Tawny red to a rich reddish brown above, tawny throat and chest with pale-brown spots on upper chest, gray speckled with even paler-brown spots below, a white throat, and a long, thin, pink down-curved bill.

Size 6.7–7.1 inches long; wingspan: 11–11.4 inches.

Behavior Males use their song to attract females. Bonding occurs over several days as male and female sing back and forth to each other. These song duets occur at dawn and at dusk.

Nest and eggs Breed north of the city. Female builds nest by constructing a platform of dead leaves on the ground under a tree or in its low branches, near a stream or in a moist, wooded area. She uses grass, stems, fibrous bark, moss, and moist, decayed leaves for the cup and lines it with mud, decayed leaves, pine needles, and rootlets, while male stands guard. Female incubates three to five glossy, pale greenish-blue eggs. Both parents feed hatchlings.

Voice Beautiful, flutelike, echoing: *day-veeur-veeur-veeur-veurrrr*, descending on the last *veeur*.

Ecological role Omnivores; feed on insects and fruit during nesting season and fruit during other seasons. When foraging for invertebrates, it uses its bill to toss leaf litter. It also gleans invertebrates from leaves. They consume fruit such as serviceberries, dogwood berries, wild cherries, blueberries, blackberries, elderberries, and sumac.

Veery perched in front of Virginia creeper vine, berry hunting, Strawberry Fields, Central Park.

Veery on porcelain berry vine, Strawberry Fields, Central Park. Though birds feed on porcelain berries, this vine is one of the most invasive plants in the Northeast, covering shrubs and trees and shading out herbaceous plants.

Hermit Thrush: *Catharus guttatus*

Where and when to find Wooded areas of city parks and backyards April and May. During fall, most common in October, sometimes lingering throughout winter.

What's in a name? *Catharus*: Latin for pure, referring to the pure notes of its song; *guttatus*: Latin for spotted, referring to its spotted breast; hermit: usually solitary; thrush: Middle English *thrusch* for this kind of bird.

Description Brown back, reddish wings and tail, white underparts with brown spots on the breast and throat. Brown head and face with a narrow white ring around the eyes.

Size 5.5–7.1 inches long; wingspan: 9.8–11.4 inches.

Behavior Forages for invertebrates in leaf litter. Uses bill to pick up dead leaves and toss them aside as it looks for moving insects and spiders. Also shakes the grass with its feet to unearth prey. Just before flight, flicks its wings or tail.

Nest and eggs Nest constructed in a depression on the ground hidden by vegetation. Female builds a cup made of weed stems and grasses with a middle layer of mud and dead leaves, lined with rootlets. She decorates the outside of the nest with lichen. Female incubates three to five glossy, pale greenish-blue eggs. Both male and female feed their young.

Voice From the very top of trees, the hermit thrush sings one of the most beautiful and haunting of all birdsongs, an echoing song with pauses between each phrase. Each song is less than two seconds long, but its beauty and purity stays with one for a long time.

Ecological role Omnivores; in spring, the hermit thrush eats mainly insects: beetles, caterpillars, bees, ants, wasps, and flies. They will also eat small amphibians and reptiles. In winter, they consume fruit, including wild berries.

Hermit thrush in holly tree. In winter, it consumes wild berries.

Hermit thrush bathing in Tanner's Spring, Central Park.

Wood Thrush: *Hylocichla mustelina*

Where and when to find Wooded areas of city parks and backyards, from late winter throughout the fall, in all five boroughs. With numbers declining due to habitat loss in its breeding grounds, it is less commonly seen.

What's in a name? *Hylocichla*: Greek for wood thrush; *mustelina*: Latin for belonging to a weasel, referring to brown and white colors; wood: habitat; thrush: Middle English *thrusch*.

Description Reddish-brown head and back; wings and tail are olive brown; white breast covered in bold, brown spots; white eye ring; white throat with spots on side of neck down to chest.

Size 7.5–8.3 inches long; wingspan: 11.8–13.4 inches.

Behavior Hops through leaf litter hunting for invertebrates by probing soil with its bill, turning leaves, gleaning invertebrates from the soil. As it forages on the ground, it gives several hops and then pauses to search for prey.

Nest and eggs Has been nesting successfully in Staten Island parks, the New York Botanical Garden, Central Park, Prospect Park, and Inwood Hill Park. Female builds nest in the lower forked branches of small trees with foliage above for protection. She begins by creating a bottom of grass, leaves, stems, and may add paper or plastic. Then she constructs grass, stems, and leafy walls and shapes a cup out of mud to incubate three to four greenish-turquoise eggs.

Voice Beautiful, haunting *tut-tut-oh-lay-oh-LEEEEEE*. This last note can be buzzy.

Ecological role Omnivores; in spring and summer, they feed on invertebrates such as beetles, flies, caterpillars, spiders, millipedes, snails, and ants. In autumn, they feed on the berries of the spicebush, blueberry, holly, elderberry, jack-in-the-pulpit, Virginia creeper, pokeweed, and fatty fruit such as dogwood and tupelo berries, which are good for all migrating birds.

Wood thrush standing on fallen log.

Wood thrush hunting invertebrates. Their foraging
pattern is to run, stop, look down for prey.

American Robin: *Turdus migratorius*

Where and when to find Year-round wherever there are trees and lawns, in frontyards and backyards, city parks, and gardens.

What's in a name? *Turdus*: a thrush; *migratorius*: wandering; *robin*: early English settlers named this bird after their robin redbreast.

Description Medium-sized bird but the largest thrush. Male: grayish brown above; head and tail almost black and the eyes have a partial white eye ring. The breast is brick red, with white outer tail feathers that are visible in flight. Bill is yellow with black tip. Females: paler than males. Juveniles have the speckled breast of other thrushes.

Size 9–11 inches long; wingspan: 14.75–16.5 inches.

Behavior Nests everywhere there are trees, and ledges with overhangs, like fire escapes. Run across lawns on long, strong legs as they hunt for worms. They cock their heads to the side, watching for movements in the grass that indicate a worm is traveling near the surface.

Nest and eggs The most successful and abundant of the city's native nesting birds; females use the wrist of one wing to build the nest's cup from dead grass, string, twigs. She plasters soft mud collected from worm castings to strengthen the nest and incubates three to seven blue-green eggs. Both parents feed the young.

Song *Cheerily, cheer up, cheer up, cheerily, cheer up.* Their syrinx (song box) has complex muscles allowing them to sing complicated songs that carry a long distance.

Ecological role Omnivores; feed on worms, beetles, grasshoppers, ants, cicadas, termites, caterpillars, butterflies, as well as holly berries, pokeberries, hawthorn berries and crab apples.

American robin collecting string for nest in April.

American robin collecting
grass for nest.

American robin nestlings
gaping for more food.

American robin trying to pull earthworm from its burrow. The worm holds on for dear life.

American robin feeding mouthful of worms to juvenile.

Female American robin on hawthorn tree in autumn.

Male American robin feeding on hawthorn berries.

UP TO 35 BIRD SPECIES of mockingbirds, thrashers, catbirds, and tremblers compose the Mimidae family: *mimid*: from Latin, to mimic. Mimids use both sides of their syrinx to produce song—sometimes they use both sides at the same time, allowing some to use two voices at once. They are renowned songsters, and some species can produce thousands of songs each day. They mimic the songs of other birds.

Mimids have long legs, long tails, and long, decurved bills. They forage in leaf litter, using their bills to toss aside debris to find invertebrates and berries. All mimids are new world birds, native to the Western Hemisphere of North and South America.

In North America, we have six mimid species: northern mockingbird, gray catbird, brown thrasher, long-billed thrasher, curve-billed thrasher, and sage thrasher.

In New York City, the northern mockingbird lives with us year-round, feeding on berries in the winter. The gray catbird and brown thrasher migrate to the city to feed and breed during spring and summer and migrate south in autumn.

GRAY CATBIRD

| GRAY | BROWN | NORTHERN |
| CATBIRD | THRASHER | MOCKINGBIRD |

Eggs shown at life size.

Gray Catbird: *Dumetella carolinensis*

Where and when to find Spring through autumn in thickets and shrubs throughout meadows, fields, and edges of city parks in all five boroughs.

What's in a name? *Dumetella*: thicket dweller; *carolinensis*: of Carolina; catbird: named for their catlike *meow* call.

Description A medium-sized gray bird with a black cap and tail; russet undertail covert feathers. Eyes and legs are black.

Size 8.3–9.4 inches long; wingspan: 8.7–11.8 inches.

Behavior Harbingers of spring, they raise their young in New York City, and in the fall migrate south where fruit, their main source of nutrition, will be available throughout the winter.

Nest and eggs Abundant nester in the city, pairs mate for life. Female constructs nest in a thicket using twigs, leaves, and bark and lines it with grasses and rootlets. Female incubates two to six turquoise-green eggs. Male feeds both female and hatchlings. On hot summer days, both parents will shade the nest with their wings. Fledglings leave nest after 11 days, but parents continue to feed them for almost two additional weeks.

Voice Part of the mockingbird family, the catbird is a great mimic, often imitating the calls of jays, hawks, and many songbirds. Rich and complicated, songs can go on for 10 minutes and include melodic whistling, gurgling, whining, and mimicry. If you hear a meowing coming from a shrub, it is most likely a gray catbird. Catbirds can use both sides of their syrinx at the same time, producing two voices simultaneously.

Ecological role Omnivores; during spring, most of their food consists of invertebrates, including ants, caterpillars, grasshoppers, beetles, spiders, and millipedes. When feeding on the ground, catbirds toss leaves aside with their bills to expose prey. As fruit becomes available in summer, it makes up half their diet. Fruits consumed include honeysuckle, grapes, dogwood, and poison ivy berries. Nestlings are fed invertebrate food almost exclusively until just before fledging, when fruit is introduced.

Juvenile gray catbird begging for food from its parent.

Gray catbird singing. Great mimics, they reproduce the
songs and calls of many birds, as well as their meow call.

Brown Thrasher: *Toxostoma rufum*

Where and when to find Coastal scrub is the favored habitat: Jamaica Bay Wildlife Refuge, coastal areas of Staten Island; also wooded edges of parks such as Central Park and Prospect Park, and backyards April through October in all five boroughs.

What's in a name? *Toxostoma*: Greek for bow mouth, referring to downcurved bill; *rufum*: Latin for red, referring to reddish-brown upper parts; *thrasher*: Old English: *thresher* meaning thrush: bird with a spotted chest.

Description Large, with long, grayish-pink legs, black, decurved bill, and brown tail that is often held up like a wren. Grayish-brown face; cinnamon-red head, back, tail, and wings, with two white wing bars. White throat; abdomen and flanks are white, covered in heavy black streaks. Eye: golden yellow.

Size 9.1–11.8 inches long; wingspan: 11.4–12.6 inches.

Behavior Moves quickly on the ground, running, walking, or hopping. Aggressively defends nest.

Nest and eggs Nests in the city in variable nesting sites: shrubs like forsythia, privet, sumac, or rose bush; lower limbs of Osage orange, cedar, elm, or honey locust trees; or on the ground. Both parents build nest using twigs, dead leaves, bark, and weed stems, and line the cup with fine grass and rootlets. Both incubate four to five pale-blue or pale-green eggs speckled all over with reddish-brown markings, making the eggs look brown. Both care for the young.

Voice One of the most accomplished singers of all birds, capable of singing thousands of songs. Brown thrashers mimic other birds, even more than mockingbirds. They emphatically, joyfully, and rapidly repeat all phrases twice from treetops.

Ecological role Omnivores; opportunistic feeders; eat invertebrates and berries. It sweeps its bill back and forth through soil and leaf litter consuming grubs, beetles, worms, crickets, sowbugs, and daddy longlegs. It plucks the berries of pokeweed, holly, hackberry trees, Virginia creeper, tupelo, bayberry, sumac, and cherry directly from trees and shrubs.

Brown thrasher at Tanner's Spring, Central Park.

Brown thrasher drinking from Tanner's Spring.

Northern Mockingbird: *Mimus polyglottos*

Where and when to find Year-round in the wooded areas of parks, gardens, and residential areas in all five boroughs.

What's in a name? *Mimus*: mimic; *polyglottos*: many tongued.

Description Medium-sized, gray above, pale below. Long, gray tail and slender black bill for plucking berries and invertebrates. White outertail feathers and white wing patches are conspicuous during flight. Dark gray stripe from bill across the eye.

Size 8.3–10.2 inches long; wingspan: 12.2–13.8 inches.

Behavior Should a dog, a cat, or any other predator go near a mockingbird nest, the male will dive-bomb the animal, driving it away from the female and nestlings. They are territorial and fierce protectors of their families and will even go after humans, hawks, and owls. A friend of mine and her husband wore hard hats from their house to their car and back, every time they passed a mockingbird nesting in a shrub next to their home.

Nest and eggs Abundant city nesters, mate for life. Each spring, they build stick nests in shrubs and understory trees and line nests with leaves, grasses, rootlets, and almost any kind of human debris—dental floss, tea bags, duct tape. They often have more than one nest at a time. The male watches over fledglings while the female incubates a new clutch of eggs. Females incubate three to five blue-green eggs with brown spots. Both parents feed the young.

Voice Known for the complicated variations of their song. On summer nights with a full moon, the male will sit in a tree and sing all night. They can sing up to 200 songs, learning new songs throughout their lives, mimicking other birds and mechanical sounds. Females are attracted to males whose songs have the greatest variety.

Ecological role Omnivores; during fall and winter, they feed mainly on berries, including holly, blackberry, pokeberry, sumac, poison ivy, rose hips, and Virginia creeper. Throughout the spring and summer, and especially during the breeding season, their diet consists of invertebrates found on lawns and meadows, which they also feed to their young.

Northern mockingbird parent feeding invertebrates to nestling.

Northern mockingbird singing from hawthorn tree. These are the greatest mimics and continue to learn new songs throughout their lives.

MEMBERS OF THE STURNIDAE family of birds, of which there are 112 species worldwide, starlings live in Europe, Asia, Africa, Australia, and the South Pacific Islands. They are old world birds. In Asia, they are called mynas, and in Africa, they are known as glossy starlings. Hildebrant's starling and the superb starling of Africa are exquisite, with iridescent purple, black, blue, and orange plumage. In Latin, *sturna* means starling.

Several starling species have been introduced to North and South America, where they have thrived. The European starling lives everywhere in North America. There are several myna species that live in Florida: the common myna and the hill myna.

Starlings are omnivores and opportunistic feeders. They are highly social and can roost in colonies of thousands of birds. When they fly together in huge numbers, it is called a murmuration, and this flight of thousands of birds sweeping up and down across the sky is amazingly beautiful.

The only member of the Sturnidae family living in New York City is the European starling, and it lives everywhere: in city parks, on city streets, in backyards, and along our rivers. They are intelligent birds and can mimic the calls of other birds, human speech, and mechanical sounds such as car alarms.

EUROPEAN STARLING

Eggs shown at life size.

European Starling: *Sturnus vulgaris*

Where and when to find Year-round throughout the five boroughs, inside and outside city parks. They are everywhere.

What's in a name? *Sturnus*: starling; *vulgaris*: common; starling: "little star," which refers to the white spots on its plumage.

Description In fall and winter, plumage is dark brown and speckled with white spots. Its bill is gray. In spring and summer, plumage is a glossy, iridescent black, showing purple and green in the sunlight, and its bill is yellow. Juveniles are a dull brown.

Size 7.9–9.1 inches long; wingspan: 12.2–15.7 inches.

Behavior Starlings walk with a waddling gait. They are fast fliers and have been clocked at more than 50 miles per hour. These intelligent birds can be taught human speech. Though there is cause for concern that they take over cavity nests of native birds, they have served humanity well. Much of what we know about bird biology is due to the study of this abundant bird.

Nest and eggs When they were released in 1890, one of their first nesting sites was under the eaves of the American Museum of Natural History, across the street from Central Park. Though they most often nest in tree cavities, city starlings also build nests on city buildings, wherever there is a cavity. The male fills the cavity with grasses, conifer needles and human litter, like string, yarn, and plastic. He uses the nest to attract a mate, who might then rearrange the nesting materials. Both parents incubate the four to eight greenish-white eggs.

Voice Starlings are mimics and have been known to imitate the songs of flickers, phoebes, crows, barking dogs, mewing cats, and mechanical sounds. In captivity, they learn human speech. In the wild, they have a wide vocal array of sounds: sharp whistles, musical warbling, chattering, trilling, and screaming.

Ecological role Omnivore; insects, including pests, such as weevils and Japanese beetles, and ants, flies, bees, wasps, grasshoppers, millipedes, snails, spiders, and earthworms. Starlings also consume fruit, berries, and seeds.

Fledgling European starling, mouth agape, screams for food from the edge of its tree cavity nest.

Parent European starling flies in with food for all.

European starling feeding its big baby on the ground.

Juvenile European starling, unbeknownst to parkgoer, waits patiently on the back of her chair at the Sheep Meadow Café for food to drop on the ground.

Adult and juvenile European starlings displaying anting behavior.

Adult European starling feeding on crab apples.

WAXWINGS belong to the family Bombicillidae. There are three species: the Bohemian waxwing, the cedar waxwing, and the Japanese waxwing. Bombicillidae is Latin for silktail. All waxwings have silky, velvety plumage, prominent head crests, black throats, and a black mask, edged in white, across their eyes. Their bodies are grayish-tawny brown with a yellow band at the end of their tail feathers. Waxwings have red, waxy looking spots on the tips of their secondary flight feathers. The red spots are caused by carotenoid pigment obtained from their diet of red fruit. Japanese waxwings nest in eastern Russia and winter in Japan, Korea, and eastern China. Bohemian waxwings live in the far northern parts of North America, Europe, and Asia. The cedar waxwings' range is only in North America.

In New York City, cedar waxwings nest near fruit trees. Some live here year-round in parks near trees that produce berries throughout the winter. They are particularly fond of cedar berries. Cedar waxwings that migrate to the city from the south, time their arrival with the ripening of sugary fruit. In summer, they also consume invertebrates, which they feed to their nestlings.

CEDAR WAXWING

Egg shown at life size.

Cedar Waxwing: *Bombycilla cedrorum*

Where and when to find Sometimes year-round feeding on fruit and berries in city parks in the five boroughs.

What's in a name? *Bombycilla*: Latin for silktail; *cedrorum*: Latin for of the cedar trees; waxwings feed on cedar berries in winter.

Description One of our most beautiful native birds. Lustrous, silky, tawny back, chest, and crest; black mask across eyes, edged in white; blue-gray wings and tail. Belly is pale yellow. Tail feathers tipped in lemon yellow. Waxy, red wing tips. Male has black throat; female brown.

Size 5.5–6.7 inches long; wingspan: 8.7–11.8 inches.

Behavior Cedar waxwings love fruit and running water. During summer, they will hawk for insects over small rivers, ponds, and lakes. If fruit is suspended from the end of a branch, they will line up and pass the fruit from bird to bird until everyone has eaten.

Nest and eggs Nests in Central Park and Prospect Park. Large nest constructed with twigs, plant down, grasses, flowers, string, and hair. Cup lined with rootlets, grasses, and conifer needles. Outside decorated with grass seeds and tree catkins. The female incubates two to six pale blue or blue-gray eggs, sometimes with black or gray spots. Hatchlings fed by both parents.

Voice A high, trilled *bzeee* and a high whistle.

Ecological role Mainly a fructivore; they feed on fruit year-round. In spring and summer, they consume serviceberries, dogwood berries, mulberries, and raspberries. In winter, they consume large amounts of cedar berries, from which they get their name, and honeysuckle, crabapple, and hawthorn berries. In summer, they can be seen hawking over water for mayflies, dragonflies, and stoneflies, catching them midair.

Cedar waxwing, showing the red, waxlike tips of its secondary feathers.

Cedar waxwing feeding on crab apples.

Female cedar waxwing (left) and juvenile (right) bathing in Tanner's Spring, Central Park. Note brown throats.

Cedar waxwings feeding on Japanese pagoda tree berries.

Male cedar waxwing showing black throat.

Cedar waxwing drinking at Tanner's Spring, Central Park.

THE NEW WORLD WARBLERS, also called wood warblers, are composed of 109 species that live in the Western Hemisphere of North, Central, and South America.

Fifty-six species live in North America. They are colorful, small, and typically arboreal with yellow, green, red, orange, blue, black, and white plumage. However, the terrestrial warblers, ovenbirds and water thrushes, are more earth colored. The active, flitting wood warblers have short, slender bills for capturing insects, but some also feed on berries and fruit. They have slender legs and slender, perching toes.

Many warblers breed in New York City or stop through in midspring to rest and feed on the way to their northern breeding grounds and again in autumn on their fall migration. Some warbler species commonly seen during spring, late summer, and early fall are the yellow-rumped, the yellow, the common yellowthroat, the black-and-white, the northern parula, the magnolia, the American redstart, the black-throated green, and the ovenbird. Central Park and most other parks offer an oasis to migrating warblers in April and May, where up to 30 different species can be seen in one day.

AMERICAN REDSTART

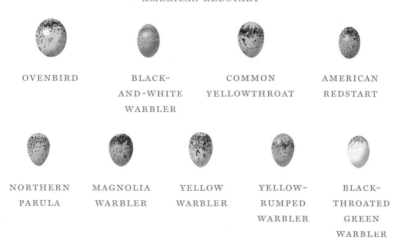

OVENBIRD

BLACK-
AND-WHITE
WARBLER

COMMON
YELLOWTHROAT

AMERICAN
REDSTART

NORTHERN
PARULA

MAGNOLIA
WARBLER

YELLOW
WARBLER

YELLOW-
RUMPED
WARBLER

BLACK-
THROATED
GREEN
WARBLER

Eggs shown at ½ life size.

Ovenbird: *Seiurus aurocapillus*

Where and when to find Spring migration: month of May; fall migration: late August through late October, in wooded areas with heavy leaf litter in city parks and backyards in all five boroughs.

What's in a name? *Seiurus*: Greek for tail wagging, which refers to its habit of jerking its tail up and slowly bringing it down; *aurocapillus*: Latin for golden crowned, referring to orange stripe on crown; ovenbird: refers to Dutch oven shape of nest.

Description Looks more like a thrush, with interesting, but less colorful plumage than a warbler, which camouflages it as it forages for invertebrates in leaf litter. Olive-brown face and head with orange crown bordered by black stripes; a greenish-brown back, black stripes alongside white throat; white underparts heavily striped with black. Male and female look alike.

Size 4.3–5.5 inches long; wingspan: 7.5–10.2 inches.

Behavior Neighboring males sing together, one starts and then others join in so that it sounds like one bird. They remember patches of leaf litter that have abundant insects and visit them year after year.

Nest and eggs Female builds a round, domed nest in a hollow on the ground, well hidden beneath leaves and branches, with a large opening on the side that cannot be seen from above, using leaves, pine needles, stems, grass, rootlets, and moss. She lines the nest with rootlets and hair and incubates four to five slightly glossy white eggs speckled with reddish-brown, brown, reddish-buff, gray, and lilac blotches that form a wreath around larger end and finer markings on the rest of the egg.

Voice *Teacher-teacher-teacher-teacher.*

Ecological role Insectivores; consume beetle adults and grubs, butterfly and moth caterpillars and adults, flies, milkweed bugs, and aphids.

Ovenbird
feeding
on moth.

Ovenbird
foraging in
leaf litter.

Black-and-white Warbler: *Mniotilta varia*

Where and when to find In city parks and wooded areas of all five boroughs. They are among the early spring migrants.

What's in a name? *Mniotilta*: moss-plucker—these warblers extract beetle larvae from moss and bark; *varia*: variegated, which refers to the black-and-white striping.

Description Small birds with lengthwise black-and-white stripes above and a white belly. They have conspicuous white stripes above and below each eye. Their legs are short, and their feet have extended hind claws for clinging to bark. Their bills, slightly curved and unusually long for a warbler, are adapted for probing bark crevices for insects. The female resembles the male but duller.

Size 4.3–5.1 inches long; wingspan: 7.1–8.7 inches.

Behavior Their short legs and long hind claws help them creep up, down, and around tree trunks and branches, searching for insects. In this regard, they are similar to nuthatches and brown creepers. They are able to arrive earlier in the spring than other warblers who consume active insects that depend on flowering plants. They feed on insects still dormant from overwintering in tree bark, using their long, curved bills to probe in and under tree bark.

Nest and eggs Nests in Riverdale Park in the Bronx. Female builds nest of grasses, dead leaves, rootlets, pine needles, and bark, typically constructed on the ground up against a tree. She incubates four to five white eggs with brown and pale purple speckles. Both parents feed the nestlings.

Voice *Weesee, weesee, weesee, weesee, weesee*: like a squeaky wheel.

Ecological role Insectivores; diet consists primarily of beetles, ants, and caterpillars, including the destructive gypsy moth caterpillars. It also feeds on spiders and daddy longlegs.

Male black-and-white warbler.

Female black-and-white warbler.

Common Yellowthroat: *Geothlypis trichas*

Where and when to find Seen from April through September in city parks in all five boroughs.

What's in a name? *Geothlypis*: ground bird; *trichas*: song thrush.

Description Tiny birds, the male has a distinctive black mask across his face, with a white band above and a yellow throat and breast. He is olive green above, with yellow undertail coverts and a white belly. The female lacks the black mask and has an olive-green face, and crown.

Size 4.3–5.1 inches long; wingspan: 5.9–7.5 inches.

Behavior Careful when they approach their nest, they do not fly directly to it, but fly to the ground and walk into the underbrush to their nest. They leave the same way, in order to keep predators from locating their nests.

Nest and eggs Nests in the city. Female constructs a loosely woven nest close to the ground in thickets and shrubs near marshes, ponds, lakes and streams, and along woody paths, using coarse plant materials for the outside layer, medium-sized plant material for the middle layer, and slender grasses, rootlets, and hair for the lining. She incubates three to six glossy, creamy-white eggs speckled with spots, scrawls and blotches of gray, reddish-brown, lilac, or black. Male feeds female and she gleans ants and small insects crawling on the nest. Both male and female feed the hatchlings.

Voice Often heard before it is seen: *witchity, witchity, witchity, witchity.* Another common call is a scolding *chack!* called as it appears from, and disappears into, dense underbrush.

Ecological role Insectivores; consume a wide variety of insects, including grasshoppers, dragonflies, damselflies, mayflies, beetles, caterpillars, moths, butterflies, flies, ants, aphids, and leafhoppers. It also feeds on spiders.

Male common yellowthroat.

Female common yellowthroat.

Juvenile male common yellowthroat.

Female common yellowthroat capturing crane flies.

Male common yellowthroat at Tanner's Spring, Central Park.

Male common yellowthroat on cotoneaster shrub.

American Redstart: *Setophaga ruticilla*

Where and when to find Throughout May; August through October, in city parks and backyards.

What's in a name? *Setophaga*: Greek for moth-eater, referring to insect diet; *ruticilla*: Latin for red tail, referring to the reddish-orange tail of the male; American: native; redstart: Anglo-Saxon *read* for red and *steort* for tail.

Description Male: black face, head, nape, back, throat, and chest; reddish-orange side patches, orange-yellow wing patches and outer tail feathers. Female: gray face, head, nape; gray to olive back, gray to white throat and chest, olive-yellow side patches, yellow outer tail feathers. Juvenile males look like females and don't develop breeding plumage until they are two.

Size 4.3–5.1 inches long; wingspan: 6.3–7.5 inches.

Behavior Flashing wing and tail patterns flushes insect prey from leaves. Their flight is an acrobatic, butterfly-like movement.

Nest and eggs Breeds in Jamaica Bay Wildlife Refuge and Ridgewood Reservoir in Queens and Mount Loretto State Preserve in Staten Island. Female builds tightly woven cup of grasses, bark, rootlets, and vine tendrils and binds it with spider webs. She lines it with fine grasses and decorates the outside with lichen, seeds, birch bark, and plant down and incubates two to five creamy-white to grayish-green eggs with dark speckles and gray spots. Bolder markings form wreath at large end.

Voice Buzzy, high pitched *wee-see, wee-see, wee-see* and *tsee-tsee-tsee-tsee-tseeee* with last note ascending.

Ecological role Insectivores; moths, beetles, flies, leafhoppers, small wasps, and spiders. During fall migration, feeds on barberries and serviceberries.

Male American redstart.

Female American redstart.

Juvenile male American redstart.

Male American redstart foraging for insects in tree.

Female American redstart with bee.

Female American redstart with potato bug.

Northern Parula: *Setophaga americana*

Where and when to find In tall "canopy" trees of city parks and woodlands and backyards, April through May and August through October as they are migrating through on their way north in the spring and on their way to forests on Caribbean Islands in autumn.

What's in a name? *Setophaga*: Greek for moth-eater, refers to its insectivorous diet; *americana*: native to the United States; *parula*: Latin for little titmouse.

Description A tiny, beautiful warbler, bluish gray above, with a green back patch; upper bill is dark gray, lower bill is orange; throat and breast yellow with chestnut and black "necklace" in-between; blackish gray in front of eye; half-moon white eye rings; two white wing bars, one short, one longer; outer tail feathers have white patches close to tip. Female lacks chestnut and black "necklace." Juveniles have less yellow on breast, a greenish cast to crown, and short, top, white wing bar.

Size 4.3–4.7 inches long; wingspan: 6.3–7.1 inches.

Behavior Feed by gleaning invertebrates from leaves and tips of twigs, flying with rapidly beating wings and hopping from branch to branch. If threatened they act hurt by drooping their wings below the base of their tails. In spring, they migrate to northern coniferous forests in bog and swamp areas that have the *Usnea*, hanging lichen to build their nests in.

Nest and eggs Female builds nest in *Usnea*, a hanging form of lichen. Using the lichen, grasses, and shredded bark, she creates a cup and lines it with plant down and rootlets. She makes an opening in the lichen at the cup level and incubates three to seven creamy white eggs, speckled with chestnut-red, purplish-red, brown, gray, and purple spots at larger end. Both parents care for the young.

Voice A rising buzzy song with a *zip, yup,* or *tsip* note at the end: *zeeeeeeeee-zip.*

Ecological role Omnivores; during warm season and when breeding, feeds on invertebrates: spiders, caterpillars, bees, wasps, grasshoppers, beetles, and moths. In its winter feeding grounds it will also eat fruit, seeds, and nectar.

Male northern parula warbler. Notice the dark "necklace" and white half-moon eye rings.

Female northern parula warbler. Note the lack of a dark "necklace."

Magnolia Warbler: *Setophaga magnolia*

Where and when to find Trees and shrubs of city parks and backyards during spring migration, April through June; and fall migration, August through early October.

What's in a name? *Setophaga*: Greek for moth-eater, referring to insect diet; *magnolia*: the tree this warbler was on in 1810 when the ornithologist Alexander Wilson named it.

Description Blue-gray crown, black back, olive shoulders and lower back; white eyebrow, black face, white half-moon lower eye ring, yellow throat and chest, with black necklace that streaks down sides. Wings grayish black with white wing patch; yellow rump, black upper tail, white under-tail coverts. Female with grayish-olive crown and nape; yellow-olive back; paler-yellow throat, chest, and abdomen with less black streaking; white eyebrow, buff half-moon lower eye ring, paler-black face.

Size 4.3–5.1 inches long; wingspan: 6.3–7.9 inches.

Behavior Gleans invertebrates from the undersides of leaves more than from leaf tops. Male sings during spring migration and when he arrives in his summer breeding ground. He also sings while nesting and raising young. Singing stops by early August.

Nest and eggs Nest is a loosely woven cup on a horizontal branch of conifer tree made of small twigs, pine needles, rootlets, and coarse grass and may be bound with spider webs. Female lines it with fine rootlets and hair, and incubates three to five creamy-white, greenish-white eggs, speckled and spotted with brown, reddish-brown, olive-tan, pale-purple and gray spots forming a wreath around larger end. Both parents care for young.

Voice Sweet, clear *weety-weety-weety-wooo* or *weety-weety-weety-weeteo* or *weeta-weeta-weetsee*.

Ecological role Insectivores; during spring migration and on breeding grounds consumes weevils, beetles, aphids, leafhoppers, caterpillars, and adult moths and butterflies. During fall migration, it may also feed on honeysuckle and Virginia creeper berries.

Male magnolia warbler in flowering black cherry tree.

Female magnolia warbler, hawking for insects.

Male magnolia warbler. Notice the black "necklace" and black streaks down his sides.

Male magnolia warbler feeding on an inchworm.

Male magnolia warbler eating winged insect.

Male magnolia warbler. Note white wing patch.

Yellow Warbler: *Setophaga petechia*

Where and when to find During spring and fall migrations in thickets, along streams, in gardens and wooded areas of city parks, in all five boroughs. May is the best time to see the yellow warbler. It nests in Van Cortlandt Park in the Bronx and Jamaica Bay Wildlife Refuge in Queens.

What's in a name? *Setophaga*: Greek for moth-eater; *petechia*: red spots on the skin.

Description All yellow from afar. On closer inspection, the wings and back are greenish yellow, and the male's breast and flanks have reddish-brown streaks. The tail has yellow spots, visible on the underside. Eyes are dark brown or black surrounded by a yellow eye ring in a yellow face. Females are paler yellow without the chestnut streaks.

Size 4.7–5.1 inches long; wingspan: 6.3–7.9 inches.

Behavior Spring migration brings these tiny, colorful, insect-eating birds to their summer breeding grounds from Central and South America where they spend the winter. Often seen in dense willow thickets near water.

Nest and eggs Female builds nest 3–12 feet off the ground usually in the fork of a willow tree. The nest is a deep cup made of bark and plant material with woven spider webs and plant down around the outside. She lines the nest with feathers and plant fibers and incubates four to five creamy white eggs speckled with gray and brown. Male feeds female as she incubates and when eggs hatch, both parents feed the young.

Voice *Sweet-sweet-sweet-I'm so sweet*, with many variations.

Ecological role Insectivores; however, occasionally will supplement diet with berries. It forages for insects and spiders on tree limbs and bushes. Small insect larvae and caterpillars are preferred foods. It feeds destructive tent and gypsy moth caterpillars among other insects to its nestlings.

Male yellow warbler in flowering apple tree.

Female yellow warbler in nettles.

Yellow-rumped Warbler: *Setophaga coronata*

Where and when to find Our most abundant warbler, found throughout the five boroughs in parks and backyards during spring and fall migration. Many overwinter by feeding on bayberries along the coast in Breezy Point, Floyd Bennett Field, Jamaica Bay Wildlife Refuge, and Jacob Riis Park in Queens, and Mount Loretto State Preserve in Staten Island. The berries of this native shrub allow them to live here year-round.

What's in a name? *Setophaga*: Greek for moth-eater, referring to its insectivorous diet; *coronata*: Latin: referring to the male's golden crown; yellowrumped: yellow feathers on its rump.

Description Yellow patches on rump, crown, and flanks. Male is blue gray above with black stripes on his back and on his chest. He has a white throat and two white wing bars, white eyebrows, and white partial eye rings below his eyes, like two little crescent moons. Female has the yellow patches on her rump and flank but is brown above, with tan eyebrows, a tan throat, and very light black markings on her chest.

Size 4.7–5.5 inches long; wingspan: 7.5–9.1 inches.

Behavior The ability to digest wax in bayberries allows them to survive winter in the northeast, which most other warblers cannot do. During warm months, it gleans insects from leaves and is known to catch them midair. During spring migration, they arrive in numbers so abundant they seem to fill the trees.

Nest and eggs Female builds nest in the north in conifer trees using grasses, twigs, rootlets, moss, lichen, and deer hair and lines cup with hair and feathers. She incubates four to five dull white eggs, speckled with dark markings around the larger end.

Voice Only male sings a bell-like trill that often ascends and speeds up at the end: *tyew-tyew-tyew-tyew-tew-tew-tew*.

Ecological role Omnivore; caterpillars, beetles, aphids, grasshoppers, flies, and spiders. During cooler months, they consume fruit: berries of juniper, poison ivy, Virginia creeper, bayberry, and dogwood and seeds of beach grass, goldenrod, sunflower, and suet from backyard bird feeders.

Male yellow-rumped warbler, showing yellow
patches on crown, chest, and rump.

Female yellow-rumped warbler
in leaf litter. Notice yellow
patches on flank and rump.

Female yellow-rumped
warbler perched on napping
sunbather's leg west of the
Great Lawn, Central Park.

Black-throated Green Warbler: *Setophaga virens*

Where and when to find Wooded areas of city parks and backyards in all five boroughs during spring migration, April through May, and fall migration, August through October.

What's in a name? *Setophaga*: Greek for moth-eater, referring to its insectivorous diet; *virens*: Greek for green, referring to their green backs; black throated: referring to the male's black throat.

Description Olive-green crown, mantle, shoulders, eye stripe, and ear patches; yellow face; bold black throat and marks descending down flanks; abdomen white; gray wings with white wing bars. Female similar but with pale yellow throat and pale black flank markings.

Size 4.3–4.7 inches long; wingspan: 6.7–7.9 inches.

Behavior Typically gleans insects from tree branches but can also hover and pluck prey from leaves.

Nest and eggs Nest is a deep cup made of twigs, moss, grass, and bark and lined with feathers and hair. Female incubates four to five grayish-white to creamy-white eggs speckled with chestnut-red, purplish-brown, purple, or gray splotches and scrawls, creating a wreath around the larger end of the egg. Both parents care for young.

Voice Typical song: thin, high-pitched, buzzy *zee-zee-zee-zee-zoo-zee and zoo-zee-zoo-zoo-zee.*

Ecological role Carnivore; consumes invertebrates, such as caterpillars, beetles, bees, moths, mites, and spiders.

Female black-throated green warbler catching insect.

Male black-throated green warbler in Tanner's Spring, Central Park.

FAMILY EMBERIZIDAE comprises buntings, sparrows, towhees, and juncos. *Embritz* is Old German for bunting. Old world Emberizids are known as buntings, and new world Emberizids are known as sparrows. All are seed-eating birds, with conical, seed-cracking bills. By some accounts, there are hundreds of species worldwide, except for Australia and Antarctica. They live in diverse habitats in temperate, tropical, and polar regions; in woods, marshes, desert, tundra, salt marshes, grassland, and bush; and in suburban and urban areas.

In North America, there are 36 species of sparrows: olive, white-colored, lark bunting, black-throated, rufous-crowned, Bachman's, Cassin's, Botteri's, chipping, clay-colored, Brewer's, vesper, field, American tree, lark, grasshopper, Baird's, Henslow's, Le Conte's, saltmarsh, Nelson's, seaside, song, Lincoln's, swamp, Savannah, fox, Harris's, white-crowned, white-throated, golden-crowned, dark-eyed junco, and towhees—spotted, eastern, green-tailed, and canyon.

In New York City, the most commonly seen species are the song sparrow, chipping sparrow, white-throated sparrow, eastern towhee, and dark-eyed junco. The sparrows and junco can be seen year-round. During mild winters, some towhees remain, but typically they migrate south in autumn and return in spring. Other sparrows that migrate into and out of the city include American tree sparrow, field sparrow, Savannah sparrow, fox sparrow, saltmarsh sharp-tailed sparrow, Nelson's sharp-tailed sparrow, Lincoln's sparrow, swamp sparrow, white-crowned sparrow.

WHITE-THROATED SPARROW

EASTERN
TOWHEE

CHIPPING
SPARROW

SONG
SPARROW

WHITE-
THROATED
SPARROW

DARK-EYED
JUNCO

Eggs shown at life size.

Eastern Towhee: *Pipilo erythrophthalmus*

Where and when to find During spring and fall migration, April through May, and October through November, along forest edges in city parks and backyards. Breeds in coastal forests in Jamaica Bay Wildlife Refuge in Queens, Floyd Bennett Field in Brooklyn, Pelham Bay Park in the Bronx, and Blue Heron Park, Goethals Pond Park, and Long Pond Park in Staten Island.

What's in a name? *Pipilo*: Latin for chirp; *erythrophthalmus*: Greek for red eyed; *towhee*: sound of its call.

Description A large, colorful, sexually dimorphic sparrow. Male has glossy black head, throat, chest, back, and tail; rufous flanks; and white abdomen. Black wings have white spots and there are long white spots on underside of tail. The female has a pale-brown head, chest, back, and tail; rufous flanks and white abdomen, with white spots on wings and under tail.

Size 6.8–8.2 inches long; wingspan: 7.9–11 inches.

Behavior Forages on the ground by kicking backward with both feet at the same time, concealed by shrubs and thickets.

Nest and eggs Nests in city. The female builds a bulky nest low in shrubs using bark, grasses, rootlets, and dead leaves and lines it with fine grasses and hair. Female incubates three to four white eggs profusely speckled with dark-brown, reddish-brown, and purplish-red spots, making the eggs look pink. Both parents care for the young.

Voice Call is *tow-hee* and song is *drink-your-tea*, with the final *tea* note stretched out in a musical trill.

Ecological role Omnivores; consume insects, spiders, millipedes, centipedes, snails, leaf and flower buds; seeds of ragweed, smartweed, and grasses; and acorns, blackberries, and blueberries.

Male eastern towhee.

Female eastern towhee.

Male eastern towhee bathing in Tanner's Spring, Central Park.

Female eastern towhee eating pin oak acorn.

Female eastern towhee in pin oak tree.

Female eastern towhee drinking at Tanner's Spring, Central Park.

Chipping Sparrow: *Spizella passerina*

Where and when to find Foraging in leaf litter and grassy lawns in city parks and backyards from April through May and August through November. Year-round in parks where they nest.

What's in a name? *Spizella*: Greek and Latin for little finch; *passerina*: Latin for sparrow.

Description Tiny, with a reddish-brown crown, white eyebrow, black eye stripe, gray cheeks, white throat; brown, tan, and white back and wings; gray rump, and pale gray breast. Colors become duller after breeding season.

Size 4.7–5.9 inches long; wingspan: 8.3 inches.

Behavior Scratches leaf litter with feet when foraging for seeds, invertebrates, and grit and will take ripe seeds from grasses within reach. Often seen feeding in small flocks on lawns and at bird feeders during fall and winter.

Nest and eggs Nests in Central Park, Prospect Park, and Green-Wood Cemetery. Female constructs loosely woven cup with dried grasses and rootlets, hidden by needles or broad leaves, at the end of southern- or eastern-facing branches to take advantage of morning sun. She lines it with fine plant fibers and animal hair and incubates two to five pale blue eggs, with black, brown, mauve, and lilac streaks, spots, or blotches, mostly at larger end of egg.

Voice A long series of very high-pitched trills made up of many monotonous chip notes, similar to a junco's song.

Ecological role Omnivores; feed mostly on grass seeds, plant material, and small fruits, except for breeding season when invertebrates are consumed. Also takes in grit to grind up seeds.

Chipping sparrow perching.

Chipping sparrow on rock.

Song Sparrow: *Melospiza melodia*

Where and when to find Year-round in city parks and backyards in all five boroughs.

What's in a name? *Melospiza*: Latin for song finch; *melodia*: Latin for pleasant song.

Description Most easily identified by heavily streaked plumage on back, chest, and flanks. Dark streaks form a central breast spot. The head has reddish-brown stripes with grayish crown and eye stripes. The tail is usually tinged with rufous-colored feathers and is fairly long. The bill is brown.

Size 4.7–6.7 inches long; wingspan: 7.1–9.4 inches.

Behavior Male sings from exposed perches on small trees as part of his courtship display or defending his territory. During ground feeding, they kick leaves aside simultaneously with both feet, as they forage for invertebrates, and catch winged termites as they emerge from underground in early June. They bathe in puddles but also bathe in drops of water on grass and leaves by beating the ground with their wings, which throws water onto their feathers.

Nest and eggs Nests throughout the city. Female builds cup-shaped nest of dried grass, bark strips, and plant stems and lines it with slender grasses, rootlets, and animal hair. She incubates two to six slightly glossy, pale-green eggs speckled with purplish-brown or reddish-brown spots. Both parents feed worms, beetles, grubs, flies, caterpillars, grasshoppers, and other insects to their nestlings.

Voice Each individual song sparrow sings from 6 to 20 different songs. The same song is often sung over and over, and then another song is sung. A common song is *fitz-fitz-fitz-weeeeee-sir-we-sir*.

Ecological role Omnivores; insects compose one-third of their diet, but weed seeds, wildflower seeds, and berries, such as serviceberries, blackberries, wild strawberries, blueberries, elderberries, raspberries, wild cherries, wild grapes, and honeysuckle berries, make up two-thirds of their diet.

Song sparrow in downy serviceberry tree. *LD*

Song sparrow on log.

White-throated Sparrow: *Zonotrichia albicollis*

Where and when to find Year-round throughout woodlands and parks of the five boroughs. They spend most daylight hours foraging among leaf litter.

What's in a name? *Zonotrichia*: striped head; *albicollis*: white neck.

Description A small, plump sparrow. Both males and females have black and white or black, brown, and tan stripes on their heads; white throats and conspicuous yellow lores between their eyes and bills. They are streaked rusty brown above and gray below. Wings are streaked rusty brown with two white wing bars.

Size 6.3–7.1 inches long; wingspan: 7.9–9.1 inches.

Behavior These sparrows are seen in small flocks as they forage together in the leaf litter, scratching backward with both feet at once. They use their bills to toss leaves aside looking for seeds, berries, and invertebrates. They also forage in the soil beneath bird feeders for sunflower and mixed seeds. Females with tan head stripes choose males with white head stripes, and females with white head stripes choose males with tan stripes.

Nest and eggs No nests have been found in the city since the 1980s. Female builds her nest on the ground or in a small shrub with grasses, roots, wood chips, and pine needles lining it with moss, rootlets, and grasses. She incubates four to six pale blue or greenish-brown eggs speckled with purple, lilac, or chestnut-red spots. Both parents feed their young.

Voice *Old Sam Peabody, Peabody, Peabody.*

Ecological role Omnivores; they consume dogwood, cedar, and spicebush berries, as well as insects, such as ants, beetles, and flies.

White-throated sparrow
feeding on elm keys.

White-throated sparrow singing. *JJ*

Dark-eyed Junco: *Junco hyemalis*

Where and when to find During winter can be found in large flocks, foraging among the leaf litter in wooded areas of city parks throughout the five boroughs.

What's in a name? *Junco*: Medieval Latin for reed bunting, which junco resembles; *hyemalis*: belonging to winter.

Description Males have a dark-gray head, back, and upper breast. The lower breast and belly are white, showing a stark contrast against the dark upper breast. Their bills are pink. Females and immatures are paler and browner than the males. Dark-eyed juncos have white outer tail feathers that are conspicuous in flight.

Size 5.5–6.3 inches long; wingspan: 7.1–9.8 inches.

Behavior Juncos scratch for food through leaf litter and soil by hopping forward and then backward with both feet at once. When there is a little bit of snow, the juncos scratch a hole 3–4 inches across to get at seeds.

Nest and eggs Nests north of the city. The male collects the nesting material and the female weaves them together. Typically built on the ground or in depressions under bushes or roots, composed of grass, leaves, and roots, lined with moss and feathers. Female incubates three to six pale-bluish or grayish-white eggs with brown, purple, and gray markings. Both parents feed hatchlings.

Voice Prolonged musical trill—*twee-wee-wee-wee-wee-wee-wee*—that can be heard hundreds of feet away. Members of a foraging flock keep in contact through a *chip* call back and forth.

Ecological role Omnivores; during winter, they forage on the ground in leaf litter and in snow for grass seeds and the seeds of ragweed, smartweed, pigweed, lamb's quarters, chickweed, sorrels, thistles, and sweetgum. During summer, they consume insects such as beetles, grasshoppers, ants, and spiders. Berries are also part of their diet.

Male dark-eyed junco in snow feeding on seeds.

Female dark-eyed junco feeding on dried berry.

CARDINALS, TANAGERS, AND GROSBEAKS

THERE ARE 66 species of birds in the Cardinalidae family, all of them in the Western Hemisphere in North America, South America, or both. They are colorful, sexually dimorphic, with males wearing brilliant plumage and females camouflaged in duller or paler feathers. The Cardinalidae family comprises cardinals, grosbeaks, tanagers, and buntings. They are seedeaters, with strong, conical seed-cracking bills. They also consume fruit and insects. Cardinalids live in wooded edges and open woodland.

In North America, there are 15 species of these colorful birds: scarlet tanager, western tanager, and summer tanager, northern cardinal, pyrrhuloxia, dicksissel, blue grosbeak, rose-breasted grosbeak, black-headed grosbeak, crimson-colored grosbeak, blue bunting, indigo bunting, lazuli bunting, painted bunting, and varied bunting.

Many of these gorgeous birds come to New York City to build their nests and raise their young, such as the scarlet tanager and rose-breasted grosbeak, or fly through during migration, such as the indigo bunting. Northern cardinals live here year-round. They are abundant in every park and area of the city with trees.

NORTHERN CARDINAL

SCARLET
TANAGER

NORTHERN
CARDINAL

ROSE-
BREASTED
GROSBEAK

Eggs shown at life size.

Scarlet Tanager: *Piranga oliveacea*

Where and when to find Wooded areas of city parks and backyards during spring migration, April through May, and fall migration, August through October.

What's in a name? *Piranga* from Tupi Indians of Amazon Basin: *tapi'rãga* for red feathers; *oliveacea:* Latin for olive green, referring to the green color of female; scarlet, referring to the scarlet color of the male; *tanager:* Tupi Indian name for this bird.

Description Male: brilliant red body, black wings, and tail. Female: greenish-yellow body, brighter yellow face and throat, olive-tan wings and tail. Nonbreeding male resembles female but has black wings and tail.

Size 6.3–6.7 inches long; wingspan: 9.8–11.4 inches.

Behavior As they forage for invertebrates, they move along tree top branches, probe the bark, hover to pluck prey off leaves and flowers, and catch bees, wasps, and hornets in midair. They swallow small invertebrates whole, but smash large prey into the tree, killing them first.

Nest and eggs Occasionally nests in Staten Island. Built at the end of a branch by the female, who constructs a shallow and loosely woven cup using twigs, rootlets, grass, and weeds and lines it with weeds, pine needles, and rootlets. She incubates two to five pale-blue eggs speckled with fine blotches of chestnut, purplish red, and gray. Both parents care for the young.

Voice From an open perch defending his territory, the male sings: *querit, queer, queery, querit, queer, querit* in his husky, two-toned, buzzy voice. Female sings a softer song in a less husky and thinner voice.

Ecological role Omnivores; eat invertebrates along with some fruit, such as mulberries and serviceberries, and new leaf buds. Their prey includes ants, sawflies, moths, butterflies, beetles, flies, cicadas, leafhoppers, spittlebugs, treehoppers, plant lice, scale insects, termites, grasshoppers, locusts, dragonflies, dobsonflies, snails, earthworms, and spiders.

Male scarlet tanager.

Female scarlet tanager devouring insect.

Nonbreeding male scarlet tanager.

Male scarlet tanager bathing, Tanner's Spring, Central Park.

Male scarlet tanager. Note thick, stocky bill for feeding on insects and fruit.

Female scarlet tanager perched in tangle of Siebold's viburnum and rose bushes.

Northern Cardinal: *Cardinalis cardinalis*

Where and when to find Year-round in our parks, backyards, street trees and shrubs.

What's in a name? *Cardinalis*: Latin: Catholic cardinal's red robe.

Description The male, except for the bit of black on his face, is a completely red bird. His thick, conical bill, crest, head, back, underparts, and tail are a brilliant red. The female is tawny above and below with a red bill and crest and reddish wings. Immature resemble females but are grayer and have darker bills. Cardinals can raise and lower their crests at will.

Size 8.3–9.1 inches long; wingspan: 9.8–12.2 inches.

Behavior They will peck at their images reflected by windows and side-view mirrors on vehicles. I saw a male cardinal pound and break all car mirrors owned by construction workers at a site. The workers replaced the mirrors and then wrapped tee shirts around them to fool the cardinal. Cardinals mate for life. When you see a cardinal, it is likely that its mate is nearby.

Nest and eggs First nested in Prospect Park in 1943, and by the 1960s, their numbers were established. Nest, constructed in shrubs or understory trees, is an open bowl of weed stems, twigs, leaves, and bark, lined with grass and often has paper or plastic hanging from the outside. During courtship, the male will feed the female. The female incubates two to five creamy eggs speckled in brown, while the male brings her food.

Voice A rich *hoit, hoit, hoit, hoit*; *what-cheer, what-cheer*; *wheat, wheat, wheat*; *pret-ty, pret-ty, pret-ty*. Their alarm call is a sharp *tsip*. Both males and females sing, often back and forth to each other.

Ecological role Omnivores; feeds on a wide variety of invertebrates, including beetles, cicadas, dragonflies, leafhoppers, aphids, ants, termites, grasshopper, crickets, spiders, snails, and slugs. It also consumes 33 kinds of wild fruit, 39 types of weed seeds, blossoms, tree seeds, and maple sap from holes made by yellow-bellied sapsuckers. A frequent visitor to bird feeders, it particularly enjoys sunflower seeds.

Male northern cardinal.

Female northern cardinal.

Female northern cardinal sitting in her well-hidden nest.

Parent northern cardinals feeding their newly fledged baby on the ground. By the next day, it had developed its tail feathers and was able to fly to a tree where the parents continued to feed it.

Father northern cardinal feeding fledgling.

Juvenile male northern cardinal.

Rose-breasted Grosbeak: *Pheucticus ludovicianus*

Where and when to find City parks, backyards, trees, and thickets during migration, throughout May and August through October.

What's in a name? *Pheucticus*: Greek for shy, secretive; *ludovicianus*: referring to Louisiana, where the bird was first named; *grosbeak*: French for large beak.

Description The male has a black head, a rose-red blaze below his black throat, a white chest and belly, a black back, and black wings with white patches. The red lining of his wings can be seen during flight. He has a thick, pale, conical seed-cracking bill. The female has a brown head with a tan stripe on the crown and above her eyes. Her throat is pale tan and her back, chest, and belly are tawny brown with dark-brown streaks. Her wings are dark brown with white patches and yellow wing linings.

Size 7.1–8.3 inches long; wingspan: 11.4–13 inches.

Behavior Both sexes sing quietly to each other when they exchange places on the nest. The male will occasionally sing inside the nest. He is a devoted mate and continues to feed his fledglings while the female builds another nest.

Nest and eggs Nests north of city. Nests are so loosely built that eggs can be seen through the bottom. Constructed of slender twigs, thick grasses, stems, and leaf litter, it is lined with twigs, hair, and rootlets. Although the male helps, the female does most of the incubating of their five pale-blue eggs with reddish-brown and reddish-purple marks that are dense around the larger end. Both parents care for the young.

Voice Song: robin-like: *cheer, cheerily, cheeryup, cheer, che-ear, che-ear, che-ea-ear*; rising and falling, musical and melodious. The male can sing hundreds of songs morning through afternoon. Call: *chink*.

Ecological role Omnivores; ants and beetles, particularly potato beetles, bees, flies, butterflies, and moths. They also feed on flowers and buds, seeds, and berries: elderberries, cotoneaster buds and berries, raspberries, mulberries, serviceberries, and smartweed, pigweed, milkweed, and sunflower seeds.

Male rose-breasted grosbeak in flowering crab apple tree.

Female rose-breasted grosbeak feeding on jewelweed fruit.

Male rose-breasted grosbeak. Notice thick, pale, conical bill, shaped for cracking seeds.

Female rose-breasted grosbeak feeding on berry.

Male rose-breasted grosbeak feeding on cotoneaster buds in May.

Male rose-breasted grosbeak's bill covered in cotoneaster flower petals.

MEMBERS OF THE ICTERIDAE family are found in the new world. More than 100 species live in the Western Hemisphere, with most living in Colombia and southern Mexico. In North America, we have bobolinks, red-winged, yellow-headed, Brewer's and rusty blackbirds, meadowlarks, common, boat-tailed, and great-tailed grackles, shiny, bronzed, and brown-headed cowbirds, and eight species of orioles: orchard, hooded, Bullock's, spot-breasted, Altamira, Audubon's, Baltimore, and Scott's.

Most icterids are black, but many have some yellow plumage; hence, the name Icterid, which in Latin means jaundiced. Icterid females are duller than the brightly colored males. Blackbird and grackle males have glossy or iridescent black feathers; some species have yellow, orange, or red markings. Icterids have large, pointed bills. Blackbirds and grackles form huge groups when they migrate. Most icterids are omnivores feeding on seeds, fruit, and invertebrates. Members of this family migrate south to their winter feeding grounds in autumn and return to their summer breeding grounds in spring.

Icterids migrate to New York City in the spring. The most commonly seen are the red-winged blackbird, the common grackle, the Baltimore oriole, and the brown-headed cowbird. Occasional visitors include the orchard oriole and the boat-tailed grackle.

RED-WINGED BLACKBIRD

RED-WINGED
BLACKBIRD

COMMON
GRACKLE

BROWN-
HEADED
COWBIRD

BALTIMORE
ORIOLE

Eggs shown at life size.

Red-winged Blackbird: *Agelaius phoeniceus*

Where and when to find March through November: commonly seen in understory trees and shrubs near marshes, ponds, lakes, and streams in city parks throughout the five boroughs.

What's in a name? *Agelaius*: belonging to a flock; *phoeniceus*: deep red.

Description Medium sized, the male is a glossy black except for his "epaulets," or shoulder patches, which are red, bordered with yellow. Sometimes the red is concealed and only the yellow is visible, until they fly, when the full, brilliant scarlet can be seen. The female is streaked brown above and below with a distinctive pale stripe above her eyes. Immature males resemble females until they are two years old, when they start to attain their adult plumage. Then they are black with red shoulder patches bordered with orange.

Size 6.7–9.1 inches long; wingspan: 12.2–15.7 inches.

Behavior Migrating males return in the spring to set up and defend their territories and attract a female. Their courtship displays and territorial defenses include spreading their tail and wings, raising their scarlet epaulet feathers, and singing their *o-ka-lee* song. The red epaulets remain covered when they want to avoid conflict with other males. However, they are fierce defenders of their nests and will mob hawks, crows, and other predators, chasing them from their territory. In the fall, they migrate south in enormous flocks.

Nest and eggs Nests all over the city. Female builds nest in wetland using cattails or reeds as vertical supports and weaves wet grasses and leaves around them. Nest is lined with mud and fine grasses. Female incubates two to six pale blue-green to gray eggs, blotched with brown streaks. Both parents feed nestlings.

Voice *O-ka-lee.*

Ecological role Omnivores; eat aquatic invertebrates, flies, beetles, moths, butterflies, gypsy moth and tent caterpillars, grubs, grasshoppers, snails, millipedes, spiders, and berries.

Adult male red-winged blackbird with red shoulder patches bordered in yellow. Juvenile males have red patches bordered in orange.

Female red-winged blackbird in its typical habitat of wetland edges.

Common Grackle: *Quiscalus quiscula*

Where and when to find City parks, woodlands, and backyards in all five boroughs year-round.

What's in a name? *Quiscalus*: Latin for quail; why this name was chosen is unknown; *common*: abundant; *grackle*: Latin *graculus* for jackdaw, or crow-like bird.

Description Long, dark bill and yellow eyes. Male has black feathers with purple, green, and blue iridescence on the head, bronze on body, and long iridescent tail. Female is smaller and duller with less iridescence. Her tail is shorter and is brown with no purple or blue. Juvenile is all brown with brown eyes.

Size 11–13.4 inches long; wingspan: 14.2–18.1 inches.

Behavior Resourceful and opportunistic feeders, they wade into water to catch small fish, snails, and tadpoles; pick leeches off the legs of turtles; and steal worms from American robins. They use the keel in their upper mandible to saw open acorns. They will spread their bodies on the ground with open wings, to let ants crawl over their feathers, a practice that is called "anting." Ants exude formic acid, a stinging chemical, which can kill parasites on the bird's body and feathers.

Nest and eggs Nests in all large city parks. Female builds nest in a tree, usually a conifer using twigs, pine needles, coarse grasses, *Usnea* lichen, and mud to make it strong. She lines it with fine grasses, rootlets, and soft material such as feathers, cloth, hair, or string. She incubates four to five glossy, pale-blue eggs spotted with bold blackish-purple or brown scrawls with fine purple-edged markings. Males help care for their young.

Voice Harsh squeaks, whistles, and croaks. Both male and female sing *readle-eak* along with high-pitched whistles.

Ecological role Omnivores; will eat what is seasonally available. In warm months: mulberries, cherries, beechnuts, invertebrate, and vertebrate animals, such as beetles, bees, grasshoppers, caterpillars, spiders, pill bugs, crabs, snails, fish, frogs, salamanders, mice, small birds, and bird eggs. In autumn, they consume acorns, weed seeds, crab apples, and hawthorn berries.

Male common grackle with feathers shimmering with blue, purple, and green iridescence.

Juvenile common grackle begging for food from its mother.

Male common grackle using the keel in his upper
mandible to saw open a pin oak acorn.

Female common grackle feeding on Hudson River mud
crab along the rocky shore of Riverside Park South.

Common grackle feeding on winged termites.

Common grackle enjoying pizza near Pier I Café, Riverside Park.

Brown-headed Cowbird: *Molothrus ater*

Where and when to find April through June and September through November in fields, meadows, and woodland edges of city parks, and in backyards.

What's in a name? *Molothrus*: Greek for greedy beggar and Latin for wanderer; *ater*: Latin for black, referring to body color; cowbird: original habitat was short-grass plains, where they foraged for insects roused by cows and bison.

Description Males: iridescent black plumage with rich brown head. Females: plain brown upper parts, with light-brown head and underparts, finely streaked belly, and dark eyes. Beak is conical, finch-like, for cracking seeds.

Size 7.5–8.7 inches long; wingspan: 14.2 inches.

Behavior Forages on the ground in mixed flocks of blackbirds. Males strut, displaying to attract females. Edge birds, they love to forage along the Belt Parkway, between the Sheepshead Bay and Cross Bay Boulevard exits.

Nest and eggs Brood parasite; female leaves her eggs in the nests of many of New York City's breeding birds. She chooses nests with eggs that are smaller than her own. The host bird incubates the cowbird's eggs and raises its young. Cowbird eggs have been laid in the nests of more than 220 species of birds. Because they do not build nests, incubate eggs, or raise their young, females expend energy laying more than three dozen brown-spotted, white eggs each season.

Voice Song: *bublo-seeleeee*: gurgling notes followed by whistles that last about a second.

Ecological role Omnivores; originally cowbirds lived in the Midwestern grasslands but have expanded their range throughout North America. They feed on grass and weed seeds throughout the year, but especially in the cold months, and invertebrates, such as beetles, snails, and grasshoppers in the summer.

Male brown-headed cowbird.

Female brown-headed cowbird, a brood parasite, lays her eggs in the nests of many of New York City's breeding birds, choosing nests with eggs smaller than her own and letting the host birds feed and raise her young.

Baltimore Oriole: *Icterus galbula*

Where and when to find Spring, summer, and fall in open wooded areas of city parks and yards throughout the five boroughs.

What's in a name? *Icterus*: jaundice (ancient Greeks believed that yellow birds cured jaundice); *galbula*: small, yellow bird; Baltimore: during the colonial era, the Baron of Baltimore's colors were orange and black.

Description The male is a brilliant orange with a black head and wings with white wing bars. Females and juveniles have a yellow face, chest, and belly, with an olive head and dark wings with two white wing bars.

Size 6.7–7.5 inches long; wingspan: 9.1–11.8 inches.

Behavior Migrate to New York City in the spring where they nest in city parks. In autumn, they return to their Central American and South American winter feeding grounds.

Nest and eggs The nest is a pouch almost 6 inches deep, and 3 to 4 inches wide at the bottom where the eggs sit, woven from plant fibers, hair, and colorful string. It hangs suspended from the tip of a branch. The females weave the nest. The inside is lined with fine grasses, feathers, and animal hair. The female incubates three to six pale, gray eggs with dark blotches. The male feeds her and the hatchlings.

Voice Males have a rich, flutelike, complex song: *chewdi-chewdi-chew-chew-che* sung from treetops to attract their mate or defend their territory during spring nesting. Each male has a unique, slightly different variation of this song. It is loud, almost tenor-like. Usually, males are so high in the tree that you hear them but have to strain your neck to see them.

Ecological role Omnivores; feed on insects, fruit, berries, and seeds; they forage in trees and shrubs for caterpillars, including tent and gypsy moth caterpillars, beetles, ants, aphids, grasshoppers, and wood borers. Its diet also includes wild cherries, serviceberries, blackberries, and grapes. It collects nectar from flowers.

Male Baltimore oriole feeding on cotoneaster buds.

Female Baltimore oriole in the cotoneaster.

Male Baltimore oriole feeding hatchlings in their hanging basket nest.

Juvenile Baltimore oriole singing.

The Baltimore oriole nest is woven from plant fibers, animal hair, and colorful string and is suspended from the tip of a branch.

Female Baltimore oriole. Orioles not only eat flowers but also feed on nectar from the flower and eat fruit and berries. Baltimore orioles are also omnivores and feed on invertebrates.

FINCHES AND OLD WORLD SPARROWS

FINCHES, FAMILY FRINGILLIDAE, Latin for songbird, and old world sparrows, family Passeridae, Latin for sparrow, are small birds with conical bills for cracking seeds. Worldwide there are 41 species of Passeridae. All are native to Europe, Africa, and Asia. Many species, like the house sparrow, were introduced by explorers and early settlers and are now completely naturalized in North America, Central America, and South America. House sparrows are also living in Iceland and on the northernmost tip of Japan on Rishiri Island. This little sparrow is the most widely distributed bird on the planet.

Throughout most of the world, the Fringillidae family comprises 140 species, consisting of finches, certain grosbeak genera, bullfinches, rosefinches, and euphonias. These birds are seedeaters with conical seed-cracking bills. The males and females are sexually dimorphous. The males are colorful; the females are more camouflaged. Finches have an undulating flight, alternating dipping and gliding with flying.

In North America, old world sparrows are represented by the house sparrows, which are everywhere. Fringillidae species include gray-crowned rosy finch, pine siskin, American goldfinch, lesser goldfinch, hoary redpoll, common redpoll, house finch, Cassin's finch, purple finch, brambling, pine grosbeak, white-winged crossbill, and red crossbill.

In New York City, house sparrows nest on every corner and in any cavity they can find: on ledges above air conditioners, in the crevices of stone walls in parks, in corner lamppost pipes in every borough. The American goldfinch and the house finch are the most commonly seen finches.

AMERICAN GOLDFINCH

| AMERICAN | HOUSE | HOUSE |
| GOLDFINCH | SPARROW | FINCH |

Eggs shown at life size.

House Finch: *Carpodacus mexicanus*

Where and when to find Year-round throughout city parks and streets in all five boroughs. In 1940, a small number of finches brought in from California for the pet trade were turned loose in Long Island. Since then they have spread across the eastern United States and southern Canada.

What's in a name? *Carpodacus:* Latin for fruit-eater; *mexicanus:* of Mexico; house: lives near and on buildings.

Description Male has a bright-red head, nape, breast, and rump. The red color of the male comes from carotenoid pigments in its food. The more pigment in the food, the redder the male. He is streaked reddish brown above, with brown and white streaking on his flanks. His wings are brown with two white wing bars. The female and immature are streaked brown, paler below, with darker wings and two white wing bars. The bill is thickly conical and downcurved. The tail is brown and square tipped, in contrast with the tail of the purple finch, which is notched.

Size 5.1–5.5 inches long; wingspan: 7.9–9.8 inches.

Nest and eggs Often nest in roof gardens of apartment buildings, in ivy-covered walls, in hanging and potted plants, under eaves, and behind air conditioners. Female prefers to mate with the reddest male she can find and proceeds to build her nest, with very little help from the male, using slender weeds, grass, fine twigs, roots and leaves, string, wool, and feathers. The inside cup of finer material is where she lays and incubates two to six pale-blue or white eggs that may be lightly speckled in blue or purple at the wider end. Her mate brings her food throughout the day, and both parents feed the hatchlings regurgitated plant material such as weed seeds and very occasionally beetle and fly larvae.

Voice A long, sweet, burbling song sung by the male that may contain some buzzy notes.

Ecological role Herbivores; consumes seeds of thistle, dandelion, ragweed, goldenrod, and fruit, such as mulberries, cherries, crab apples, and hawthorn berries. House finches are frequent visitors to bird feeders.

Male house finch.

Female house finch feeding on crab apples.

Male feeding juvenile house finch.

Male house finch collecting mulberries while juvenile waits.

Male house finch feeding on dandelion seeds.

Male house finch at water's edge.

American Goldfinch: *Carduelis tristis*

Where and when to find Year-round at backyard feeders and throughout city parks in all five boroughs. To attract American goldfinches, plant thistles or put out thistle seed in your bird feeders.

What's in a name? *Carduelis:* thistle finch; *tristis:* sad (in reference to its call).

Description During the spring and summer breeding seasons, the male is a small, bright-yellow bird with a black cap and forehead, white rump, black wings, and a black tail with white edges. Females are a dull, pale yellowish tan with black wings, black tail, and white wing bars. Males resemble females in winter.

Size 4.3–5.5 inches long; wingspan: 7.5–8.7 inches.

Behavior Forage for seeds in mixed flocks in winter and small flocks in summer.

Nest and eggs Nests in city. Nests are made of woven plant fibers, lined with the down of thistle and milkweed. They are so thick, they can temporarily hold water. Spider silk and caterpillar webs along with bark and strong plant fibers such as grape or hawthorn are used to bind the rim. Females do most of the gathering and construction, incubating up to seven white eggs.

Voice Their soprano song *per-chick-o-ree* is often repeated at each rise of their undulating flight.

Ecological role Herbivores; feeds on seeds of wildflowers, weeds, and trees. Thistle seeds are their favorite, but they can often be seen feeding on the seeds of dandelion, sunflower, chicory, ragweed, goldenrod, and a variety of grasses.

Male American goldfinch in breeding plumage.

Female American goldfinch, nonbreeding plumage,
feeding on tiny seeds in sweetgum tree seed pod.

House Sparrow: *Passer domesticus*

Where and when to find One hundred house sparrows were introduced from Europe into Brooklyn, Manhattan, and Chicago in the early 1850s, and the species expanded throughout North America. It is the most commonly seen bird throughout the five boroughs and lives here year-round.

What's in a name? *Passer:* Latin for small, active; *domesticus:* Latin for belonging to a house.

Description The male has a gray crown with chestnut patches bordering the crown and extending down to the pale-gray cheek and neck. The black stripe in front of the eye extends to the beak and meets the black bib. The thick bill is grayish black, and the legs are pale brown. The rump and tail are gray, the shoulders are chestnut brown, and the wings are brownish with a white wing bar. The female has light brown cheeks, neck, and breast without the black bib. She has a buffy stripe between a brown eye stripe and brown crown.

Size 5.9–6.7 inches long; wingspan: 7.5–9.8 inches.

Behavior House sparrows take baths in patches of dusty soil, usually in large groups. Each little bird creates a depression and throws dust all over its feathers to destroy parasites.

Nest and eggs House sparrows mate for life. They live and nest inside every street corner lamppost pipe, over air conditioners, and inside any cavity they can find on building exteriors, dock pilings, and window grates. Within these cavities, they construct their nests with dried grasses, feathers, and string. I have seen them emerge from places that are surprising, like the nostrils of Teddy Roosevelt's bronze horse on Central Park West in front of the American Museum of Natural History.

Voice A series of *cheeps* and *chirrups.*

Ecological role House sparrows are omnivores that feed on fruit in summer and dried berries and grass seeds in winter. In summer, they consume invertebrates: beetles, cicadas, grasshoppers, crickets, aphids, spiders, flies, and moths. Year-round they feed on human food litter on the ground.

House sparrow nest in corner traffic signal pole.
Look for them on almost every corner.

Female house sparrow feeding her nestlings,
with bright-yellow gapes and bright red mouths.

Male house sparrow collecting nesting material.

Male house sparrow feeding on Japanese beetles.

Female house sparrow feeding fledgling.

Male house sparrow consuming winged termites.

BIRDING ORGANIZATIONS
AND RESOURCES

Organizations

New York City Audubon
http://www.nycaudubon.org/
Through workshops, bird walks, trips, classes, newsletters for adults
and children, and conservation programs such as Project Safe Flight,
Harbor Herons, Jamaica Bay Project, and Lights Out New York, this
wonderful organization lives up to its mission to protect wild birds and
their habitats throughout the five boroughs, thus improving the quality
of life for all New Yorkers.

American Littoral Society, Northeast Region
http://www.littoralsociety.org/index.php/chapters1/northeast
-chapter/jamaica-bay-guardian
The American Littoral Society's mission is to protect marine habitat
and its wildlife from harm and to inspire others to work toward this
goal through education, field trips, native plantings, and volunteerism.
The Northeast Region is run by Don Riepe, Guardian of Jamaica Bay
Wildlife Refuge.

Linnaean Society of New York
http://linnaeannewyork.org/
Meetings of the Linnaean Society of New York are open to the public
at the American Museum of Natural History on the second Tuesday of
every month, from September through May, except in March. Scientists
and naturalists give illustrated talks, and birding field trips are held
throughout the city parks for members and are open to nonmembers,
space permitting.

American Museum of Natural History

http://www.amnh.org/learn-teach/adults/nature-walks

Museum ornithologists Paul Sweet and Joseph DiConstanzo lead bird walks and nature walks.

Brooklyn Bird Club

www.brooklynbirdclub.org/contacts.htm

Birding field trips in the five boroughs led by knowledgeable and passionate birders from this century-old bird club founded in 1909 whose mission has been to promote birding and bird habitat conservation in Brooklyn and beyond.

City Island Birds

www.cityislandbirds.com

Jack Rothman and other birders lead bird walks in Pelham Bay Park in the Bronx.

Queens County Bird Club

www.qcbirdclub.org

Founded in 1932, with two conditions: "a person must have a profound interest in bird life and must be at least 12 years of age."

Staten Island Museum

www.statenislandmuseum.org

Offers bird walks throughout Staten Island's many wonderful inland and coastal parks.

Urban Park Rangers

www.nycgovparks.org/programs/rangers

The rangers lead terrific nature walks throughout city parks.

Fort Tryon Park

www.forttryonparktrust.org

Experienced naturalists lead bird and nature walks, free to the public, through this gorgeous park high over the banks of the Hudson River.

New York City Department of Parks and Recreation
http://www.nycgovparks.org/events
New York City Department of Parks and Recreation posts outdoor
nature events on this website that are taking place in its parks throughout the five boroughs.

The Nature Conservancy
http://www.nature.org/ourinitiatives/regions/northamerica/united-
states/newyork/placesweprotect/newyorkcity/
Since 1995, The Nature Conservancy's LEAF program—Leaders in
Environmental Action for the Future—has recruited teenagers from
New York City high schools and provided them with paid summer
internships in nature preserves to restore wildlife habitat, to learn
about wildlife, and to expose them to career possibilities.

Rehabilitators

*Without these men and women, hurt, orphaned, and sick wildlife would
suffer even more. These are the people who tirelessly, year after year, open
their hearts, their wallets, and their homes to birds and other wildlife to
heal them and, if possible, release them back into the wild.*

Bobby and Cathy Horvath
Wildlife in Need of Rescue and Rehabilitation
https://www.facebook.com/pages/WINORR-Wildlife-In-Need-of
-Rescue-and-Rehabilitation/113685721999067?sk=info
202 N. Wyoming Avenue, N. Massapequa, NY 11758
516-293-0587
Nonprofit volunteer organization that provides professional care for
sick, injured, and orphaned wildlife.

Rita McMahon
The Wild Bird Fund, http://www.wildbirdfund.org
565 Columbus Avenue (between 87th and 88th Streets)
New York City, NY 10024
646-306-2862

Provides medical care and rehabilitation to injured, sick, and orphaned New York City wildlife.

Urban Park Rangers
http://www.nycgovparks.org/programs/rangers
Call 311 and ask for Urban Park Rangers, or call main office: 212-360-2774
Will rescue birds in New York City parks.
Ask for Ranger Rob Mastrianni, who helps birds of prey.

Photographers' Blogs

The passionate men and women who capture the beauty of New York City birds throughout the season for all to see simultaneously monitor the birds' health and the health of their habitat. Their tireless work has alerted New Yorkers when a bird's life is at risk due to poisoning or when they look sick or injured. They are the eyes and voice of the birding community, as they are outside every day, in every borough, every season, year in and year out. New Yorkers who care about our wildlife and their environment can go online and see which bird species have migrated into our city, what they are eating, where they are bathing, and how they are caring for their young. These photographers and their blogs are a great gift to our city.

Andrew Baksh: birdingdude.blogspot.com

Beth Bergman: http://thebethlenz.blogspot.com

Bob DeCandido: Birdingbob.com

Joe DiCostanzo: inwoodbirder.blogspot.com

Peter Dorosch: http://prospectsightings.blogspot.com

Phil Jeffrey: nycbirding.blogspot.com

Rob Jett: citybirder.blogspot.com

Lincoln Karim: www.palemale.com

Steve Nanz: http://www.stevenanz.com/

James O'Brien: http://yojimbot.blogspot.com

Francois Portmann: http://www.fotoportmann.com/birds/

Joe Reynolds: http://natureontheedgenyc.blogspot.com

Robert: http://morningsidehawks.blogspot.com

Jean Shum: www.jeanshum.com

David Speiser: http://www.lilibirds.com/gallery2/main.php

Lloyd Spitalnick: http://www.lloydspitalnikphotos.com/

Marie Winn: http://mariewinnnaturenews.blogspot.com/

D. Bruce Yolton: http://urbanhawks.blogs.com

https://twitter.com/BrooklynBirder/brooklyn-bird-sightings

Tracking Birds

American Bird Conservancy: www.abcbirds.org

Ebird.org

Migratory Bird Research: www.mbr-pwrc.usgs.gov

North American Rare Bird Alert: www.narba.org

http://www.nycbirdreport.com

Online Bird Sites

Bird calls: www.xeno-canto.org/

Cornell Lab of Ornithology: www.allaboutbirds.org; http://bna.birds
.cornell.edu

New York State and National Organizations

American Birding Association: www.americanbirding.org

Hawk Migration Association of North America: www.hmana.org

National Audubon Society: www.audubon.org

Natural Resources Defense Council: www.nrdc.org/

The Nature Conservancy: www.tnc.org

New York State Ornithological Association: www.nybirds.org/

New York State Young Birders Club: www.nysyoungbirders.org

BIRDING HOTSPOTS

Birding hotspots in New York City and surrounding neighborhoods were compiled from ebird.org.

Manhattan Sites and Surrounding Neighborhoods

Battery Park North Cove and South Cove: Battery Park City

Bryant Park: Midtown, Theater District, Garment District, Murray Hill

Central Park: Midtown West, Lenox Hill, Upper West Side, Upper East Side, Carnegie Hill, Spanish Harlem, Manhattan Valley, Morningside Heights, Harlem

Fort Tryon Park: Washington Heights, Inwood

Four Freedoms and Park: Roosevelt Island

Governor's Island

Greenwich Village Waterfront: West Houston to 14th Street

Highbridge Park: Washington Heights, Inwood

Inwood Hill Park: Inwood, Washington Heights

Madison Square Park: Flatiron District, NoMad

Marcus Garvey Park: Harlem

Morningside Park: Harlem, Morningside Heights

Peter Detmold Park: Turtle Bay Gardens

Randall's Island

Riverbank State Park: Hamilton Heights, Sugar Hill (Harlem)

Riverside Park: Upper West Side, Manhattan Valley, Harlem

Theodore Roosevelt Park: Upper West Side, American Museum of Natural History

South Street Seaport: Financial District

Stuyvesant Cove Park: Stuyvesant Town, Gramercy

Swindler Cove Park and Sherman Creek: Inwood

Union Square Park: Greenwich Village, East Village, Flatiron District, Gramercy

Washington Square Park: East Village, West Village

Bronx Sites and Surrounding Neighborhoods

Barretto Point Park: Hunts Point

Bronx Zoo: East Tremont, Fordham, Belmont, Van Nest, Morris Park

City Island

Clason Point Park: Clason Point, Soundview

New York Botanical Garden: Bedford Park, Fordham, Belmont, Morris Park, Norwood

Orchard Beach: Coop City, Eastchester, Country Club

Pelham Bay Park: Coop City, Eastchester, Baychester, Pelham Gardens, Country Club

Pelham Bay Park South: Pelham Bay, Middletown, Spencer Estates and Country Club

Pugsley Creek Park: Clason Point, Castle Hill

Riverdale Park: Riverdale, Fieldston, Kingsbridge, Marble Hill

Soundview Park: Clason Point, Soundview

SUNY Maritime College, Fort Schuyler: Throgs Neck

Van Cortlandt Park: Fieldston, Riverdale, Kingsbridge

Wave Hill: Riverdale, Fieldston

Williamsbridge Reservoir: Norwood

Woodlawn Cemetery: Norwood

Brooklyn Sites and Surrounding Neighborhoods

Brooklyn Botanic Garden: Park Slope, Prospect Lefferts Gardens, Prospect Heights

Brooklyn Bridge Park: Brooklyn Heights, Vinegar Hill, DUMBO (Down Under the Manhattan Bridge Overpass)

Canarsie Pier: Canarsie

Coney Island Creek Park: Seagate

Dead Horse Bay: Mill Basin, Gerritsen Beach

Dreier-Offerman Park / Calvert Vaux Park: Gravesend

Floyd Bennett Field: Mill Basin, Bergen Beach, Marine Park, Gerritsen Beach

Fort Green Park: Fort Green, Downtown Brooklyn

Fresh Creek Park: Canarsie, East New York

Gateway National Recreation Area: Bergen Beach, Mill Basin

Gravesend Bay: Gravesend

Green-Wood Cemetery: Windsor Terrace, Kensington, South Slope, Sunset Park

Hendrix Creek: East New York

Marine Park Salt Marsh: Marine Park, Gerritsen Beach, Mill Basin

Owls Head Park: Bay Ridge, Sunset Park

Paerdegat Basin: Canarsie, Bergen Beach

Pier 44 Waterfront Garden: Redhook

Plumb Beach: Sheepshead Bay

Prospect Park: Park Slope, Prospect Heights, Windsor Terrace, Prospect Lefferts Gardens, Crown Heights, Prospect Park South

Ridgewood Reservoir (also part of Queens): Cypress Hills

Sheepshead Bay: Sheepshead Bay, Manhattan Beach, Brighton Beach

Veteran's Memorial Pier: Bay Ridge

Queens Sites and Surrounding Neighborhoods

Alley Pond Park: Bayside, Douglaston, Little Neck, Bellerose

Baisley Pond Park: Springfield Gardens, St. Albans

Big Egg Marsh: Broad Channel

Breezy Point Tip: Breezy Point, The Rockaways

Brookville Park: Rosedale, Brookville

Cunningham Park: Hollis Hills, Fresh Meadows

Dubos Point: Arverne, Rockaway Beach

Edgemere Landfill: Edgemere

Flushing Meadows Corona Park: Corona, Forest Hills, Kew Gardens

Forest Park: Forest Hills, Woodhaven, Ridgewood, Ozone Park, Richmond Hill

Fort Tilden Beach and hawk watch platform: Breezy Point, Rockaway Park, Roxbury

Fort Totten Park: Bayside

Highland Park: Cypress Hills and Bushwick (Brooklyn), Ridgewood, Woodhaven

Jacob Riis Park: Rockaway Park, Breezy Point, Roxbury

Jamaica Bay Wildlife Refuge: Broad Channel, Howard Beach, Rockaway Park

Kissena Park: Flushing, Utopia

Little Neck Bay: Douglaston, Bayside

Queens Botanical Garden: Flushing

Ridgewood Reservoir (also part of Brooklyn): Ridgewood, Woodhaven

Rockaway Park: Rockaway Beach, Belle Harbor

World's Fair Marina: Astoria, Flushing, Elmhurst

Staten Island

Acme Pond: Annadale, Prince's Bay

Arbutus Lake: Huguenot

Arden Avenue Beach / Mayberry Promenade: Southeast Annadale

Bloomingdale Park: Woodrow, Charleston

Blue Heron Park: Annadale

Bucks Hollow Greenbelt: New Springville, Lighthouse Hill

Cemetery of the Resurrection: Prince's Bay

Clay Pit Ponds: Charleston

Clove Lakes Park: Castleton Corners

College of Staten Island: New Springville, Manor Heights, Bulls Head

Conference House Park: Tottenville

Faber Park: Port Richmond

Franklin D. Roosevelt Boardwalk and Beach: South Beach

Freshkills Park: Greenridge, Arden Heights

Goethals Bridge Park and Creek: Bloomfield/Arlington

Great Kills / Gateway Park: Great Kills

High Rock Park: Egbertville / Lighthouse Hill

Kingfisher Pond: Great Kills

Lemon Creek Pier: Prince's Bay

Mariner's Marsh: Mariner's Harbor

Mill Creek: Bulls Head, Bloomfield, Travis-Chelsea

Miller Field: New Dorp, Midland Beach / New Dorp Beach

Moravian Cemetery: New Dorp

Mount Loretto State Preserve: Prince's Bay

Oakwood Beach: Oakwood

Ocean Breeze Fishing Pier: Ocean Breeze / Midland Beach

Page Avenue Beach: Tottenville

Saw Mill Creek Marsh: Bloomfield

Seguine Pond and waterfront: Annadale

Sharrotts Shoreline: Charleston

Silver Lake Park: Silver Lake

Snug Harbor / Staten Island Botanical Garden: Randall Manor /
    New Brighton

Wagner College: Grymes Hill

William T. Davis Wildlife Refuge: New Springville

Willowbrook Park: Willowbrook / Bulls Head

Wolfe's Pond Park: Huguenot

BIBLIOGRAPHY

Baicich, Paul J., and Colin J. O. Harrison. *Nests, Eggs, and Nestlings of North American Birds*. Princeton, NJ: Princeton University Press, 1997.

Bull, John. *Birds of the New York Area*. New York: Dover, 1964.

Day, Leslie. *Field Guide to the Natural World of New York City*. Baltimore: Johns Hopkins University Press, 2007.

Fowle, Marcia T., and Paul Kerlinger. *The New York City Audubon Society Guide to Finding Birds in the Metropolitan Area*. Ithaca, NY: Comstock, 2001.

Holloway, Joel Ellis. *Dictionary of Birds of the United States*. Portland, OR: Timber Press, 2003.

Levine, Emanuel, ed. *Bulls Birds of New York State*. Ithaca, NY: Comstock, 1998.

Martin, Alexander C., Herbert S. Zim, and Arnold L. Nelson. *American Wildlife & Plants: A Guide to Wildlife Food Habits*. New York: Dover, 2011.

Matthews, F. Schuyler. *Field Book of Wild Birds and Their Music*, 15th ed. Bedford, MA: Applewood Books, 2001.

McGowan, Kevin J., and Kimberley Corwin, eds. *The Second Atlas of Breeding Birds in New York State*. Ithaca, NY: Comstock, 2008.

Peterson, Roger Tory. *Peterson Field Guides to Eastern Birds*, 4th ed. Boston: Houghton Mifflin, 1980.

Sibley, David Allen. *The Sibley Field Guide to Birds*. New York: Alfred A. Knopf, 2003.

Terres, John K. *The Audubon Society Encyclopedia of North American Birds*. New York: Wings Books, 1991.

PHOTOGRAPHER CREDITS

All photos are by Beth Bergman unless the legend
is followed by the photographer's initials.

DBY	D. BRUCE YOLTON
DG	DAVID GOLDEMBERG
DR	DON RIEPE
JM	USDA WILDLIFE SERVICE PHOTOS BY JENNY MASTANTUONO
JO	JAMES O'BRIEN
LD	LESLIE DAY
LM	LAURA MEYERS
MP	MARGE PANGIONE
SS	SAMMIE SMITH

INDEX

*The letter "f" following a page number indicated a figure;
a "pl" indicates a plate.*

NOTES

NOTES

NOTES

Inspired since childhood by nature and the vibrancy of her city, **Leslie Day** is the author of *Field Guide to the Natural World of New York City* and *Field Guide to the Street Trees of New York City*. Her interest in neighborhood birds harkens back to a female cardinal that accompanied her on her walks for years. Interest in that cardinal fostered a lifelong pursuit of birding. Once a resident of the 79th Street Boat Basin, Day came to know waterfowl as well as the city's songbirds. Passionate about science, she earned a doctorate in science education from Columbia University and worked for decades as a science teacher. Now retired from teaching, Day spends her time writing books and leading nature walks throughout the city for organizations such as the New York Historical Society, the High Line Park, Fort Tryon Park Trust, and the New York City Department of Parks and Recreation. She visits classrooms throughout the City's schools and can be found lecturing on the wonders of New York's natural world to senior citizens.

A lover of nature and language, **Trudy Smoke** is a professor in the Department of English at Hunter College. She earned her PhD in linguistics and rhetoric from New York University and studied drawing at the New York Botanical Garden. Smoke illustrated Leslie Day's *Field Guide to the Street Trees of New York City*. Twenty-one of her leaf drawings are on display in the New Leaf Restaurant in Fort Tryon Park, New York. Her artistic style is reminiscent of the works of Mark Catesby and John James Audubon.

For more than four decades, **Beth Bergman** has been regarded as one of the world's premier opera photographers. Her photographs in this book reflect a second passion, nature. Since childhood, Beth has had a keen interest in birds and other wildlife, and her photographs of them have been popularized in her blog, thebethlenz. Beth's photographs have appeared in books, recordings, private collections, online, and even a Canadian postage stamp.